AF388674

Optoelectronic Materials, Devices, and Applications

Optoelectronic Materials, Devices, and Applications

Editors

Pingjuan Niu
Li Pei
Yunhui Mei
Hua Bai
Jia Shi

Basel • Beijing • Wuhan • Barcelona • Belgrade • Novi Sad • Cluj • Manchester

Editors

Pingjuan Niu
Tiangong University
Tianjin, China

Li Pei
Beijing Jiaotong University
Beijing, China

Yunhui Mei
Tiangong University
Tianjin, China

Hua Bai
Tiangong University
Tianjin, China

Jia Shi
Tiangong University
Tianjin, China

Editorial Office
MDPI
St. Alban-Anlage 66
4052 Basel, Switzerland

This is a reprint of articles from the Special Issue published online in the open access journal *Applied Sciences* (ISSN 2076-3417) (available at: https://www.mdpi.com/journal/applsci/special_issues/Optoelectronic_Materials_Devices_Applications).

For citation purposes, cite each article independently as indicated on the article page online and as indicated below:

Lastname, A.A.; Lastname, B.B. Article Title. *Journal Name* **Year**, *Volume Number*, Page Range.

ISBN 978-3-0365-8648-9 (Hbk)
ISBN 978-3-0365-8649-6 (PDF)
doi.org/10.3390/books978-3-0365-8649-6

Contents

About the Editors

Pingjuan Niu

Pingjuan Niu was received the B.Eng. and M.S. degrees from the Hebei University of Technology, Hebei, China, in 1996 and 1999, respectively, and the Ph.D. degree in microelectronics and solid state electronics from Tianjin University, Tianjin, China, in 2002. She is currently a Professor with Tiangong University, Tianjin. Her current research interests include semiconductor devices, power electronic devices, and intelligent sensors.

Li Pei

Li Pei is currently a Professor of Beijing Jiaotong University, her current research interests include high speed optical telecommunication network, optical fiber sensor, key technology of optical fiber communication and so on. She has been a leader of more than 10 projects in China, including National Natural Science Funds for Distinguished Young Scholar, Hi-Tech research and Development Program of China, New Century Excellent Talents Project of Ministry of Education, and so on. She has authored/co-authored more than 200 journal and conference papers, and more than 100 patents and software copyrights in her research field.

Yunhui Mei

Yunhui Mei is currently a Professor with the School of Electrical Engineering, Tiangong University, Tianjin, China. From 2011 to 2020, he was a Professor with the School of Material Science and Engineering, Tianjin University. He was with the Center for Power Electronics Systems (CPES), Virginia Tech, Blacksburg, VA, USA. He has authored more than 130 papers and 25 granted patents on power electronic packaging. His current research interests include power packaging, materials, and reliability for high power-density and high temperature applications. He was the recipient of the 2016 IEEE CPMT Japan Chapter Young Award, and 2017 IEEE APEC Outstanding Presentation Award. He was granted as the Outstanding Youth Scientist of NSFC. Dr. Mei, was recipient of the 2016 IEEE CPMT Japan Chapter Young Award, and 2017 IEEE APEC Outstanding Presentation Award, and so on. He was also the recipient of the first-class Technical Innovation Award of China Power Supply Society and China Electrotechnical Society in 2019 and 2020. He was granted as the Outstanding Youth Scientist of NSFC. He was granted as Outstanding Youth Award of Natural Science Foundation of China in 2019, and Distinguished Young Award of Tianjin Municipal Science and Technology Bureau in 2021.

Hua Bai

Hua Bai received the B.S. degree, M.S. degree, and Ph.D. degrees in 2003, 2007, and 2011, respectively, from Nankai University, Tianjin, China. He is currently a Professor with the School of Electrical Engineering, Tiangong University. His research interests include signal processing, photoelectric detection technology and systems, and biophotonics.

Jia Shi

Jia Shi was born in Jiangsu Province, China, in 1990. He received the B.S. degree in electronic information science and technology from Tiangong University, Tianjin, China, in 2013, and the Ph.D. degree in optoelectronics and photonics technology from Tianjin University, Tianjin, China, in 2018. He joined Tiangong University in China as a Lecturer in 2018. He is currently an Associate Professor with Tiangong University. His research interests include optical fiber sensors and terahertz technology.

Preface

This book, sponsored by the Special Issue entitled "Optoelectronic Materials, Devices, and Applications", is devoted to gathering a broad array of research papers on the latest advances in the development of optoelectronic materials and devices of semiconductors, fiber optics, power electronics, microwaves, and terahertz. Each of the included papers highlights the latest principles, methods, and potential applications of optoelectronics. The primary aim of this Special Issue is to promote cross-disciplinary research in optoelectronics. It is our sincere hope that these advances will provide new inspiration for the development and application of optoelectronic materials and devices. We would like to give our thanks to all of the authors and peer reviewers who contributed to this Special Issue.

Pingjuan Niu , Li Pei, Yunhui Mei, Hua Bai, and Jia Shi
Editors

applied sciences

Editorial

Optoelectronic Materials, Devices, and Applications

Pingjuan Niu [1,*], Li Pei [2], Yunhui Mei [1], Hua Bai [1] and Jia Shi [1]

[1] Tianjin Key Laboratory of Optoelectronic Detection Technology and System, School of Electronic and Information Engineering, Tiangong University, Tianjin 300387, China; meiyunhui@tiangong.edu.cn (Y.M.); baihua@tiangong.edu.cn (H.B.); shijia@tiangong.edu.cn (J.S.)

[2] Institute of Lightwave Technology, Beijing Jiaotong University, Beijing 100044, China; lipei@bjtu.edu.cn

* Correspondence: niupingjuan@tiangong.edu.cn

Citation: Niu, P.; Pei, L.; Mei, Y.; Bai, H.; Shi, J. Optoelectronic Materials, Devices, and Applications. *Appl. Sci.* **2023**, *13*, 7514. https://doi.org/10.3390/app13137514

Received: 19 June 2023

Accepted: 21 June 2023

Published: 25 June 2023

This Special Issue entitled "Optoelectronic Materials, Devices, and Applications" is devoted to gathering a broad array of research papers on the latest advances in the development of optoelectronic materials and devices of semiconductors, fiber optics, power electronics, microwaves, and terahertz. Each of the included papers highlights the latest principles, methods, and potential applications of optoelectronics. The primary aim of this Special Issue is to promote cross-disciplinary research in optoelectronics.

In total, ten papers are included in this Special Issue. New advances in optoelectronic materials have been reported for crystals, electrodes, and bonding materials. First, Zakrzewski et al. analyzed the photothermal piezoelectric spectroscopy of $Cd_{1-x}Be_xTe$, a new material with potential for use in X-ray and γ-ray detectors [1]. Next, Han et al. investigated the characteristics of a copper foil three-electrode planar spark gap high-voltage switch integrated with EFI [2]. Ding et al. showed in their study a reliable way to improve the electrochemical migration resistance of nanosilver paste as a bonding material [3]. New advances in optoelectronic devices have also been reported for LEDs and photonic crystal waveguides. Bai et al., for example, proposed a new method for the measurement of adhesive force between a single μLED and a substrate based on the use of an atomic force microscope [4]. Zhang et al. analyzed the strain relaxation effect on the peak wavelength of blue InGaN/GaN multi-quantum well micro-LEDs [5]. In addition, Shi et al. proposed an all-dielectric terahertz photonic crystal waveguide with a lilac-shaped defect operating in a 6G terahertz communication window [6]. New applications of optoelectronic materials and devices have additionally been reported for piezoelectric sensors, crystal materials, synthetic aperture radar (SAR), and optical coherence tomography (OCT). Wang et al. established a collision model of wheat grains impacting a force plate with a piezoelectric sensor and investigated the influence of the elastic recovery coefficient on the sensor's detection accuracy during the collision process [7]. Next, He et al. demonstrated the application of a $BaGa_4Se_7$ crystal in a tunable and compact mid infrared optical parametric oscillator with a repetition rate of up to 250 Hz [8]. Wang et al. proposed a new feature learning method for the automatic target recognition of SAR images [9]. Finally, Shi et al. reviewed the quantitative assessment methods used for early enamel caries with OCT [10].

It is our sincere hope that these advances will provide new inspiration for the development and application of optoelectronic materials and devices.

Acknowledgments: We would like to give our thanks to all of the authors and peer reviewers who contributed to this Special Issue entitled "Optoelectronic Materials, Devices, and Applications".

Conflicts of Interest: The authors declare no conflict of interest.

References

1. Zakrzewski, J.; Strzałkowski, K.; Boumhamdi, M.; Marasek, A.; Abouais, A.; Kamiński, D.M. Photothermal Determination of the Surface Treatment of Cd1-xBexTe Mixed Crystals. *Appl. Sci.* **2023**, *13*, 2113. [CrossRef]
2. Han, K.; Zhao, W.; Deng, P.; Chu, E.; Jiao, Q. Research on Characteristics of Copper Foil Three-Electrode Planar Spark Gap High Voltage Switch Integrated with EFI. *Appl. Sci.* **2022**, *12*, 1989. [CrossRef]
3. Ding, Z.; Wang, Z.; Zhang, B.; Lu, G.-Q.; Mei, Y.-H. A Reliable Way to Improve Electrochemical Migration (ECM) Resistance of Nanosilver Paste as a Bonding Material. *Appl. Sci.* **2022**, *12*, 4748. [CrossRef]
4. Bai, J.; Niu, P.; Cao, S.; Liu, Q. The Adhesive Force Measurement between Single μLED and Substrate Based on Atomic Force Microscope. *Appl. Sci.* **2022**, *12*, 9480. [CrossRef]
5. Zhang, C.; Gao, K.; Wang, F.; Chen, Z.; Shields, P.; Lee, S.; Wang, Y.; Zhang, D.; Liu, H.; Niu, P. Strain Relaxation Effect on the Peak Wavelength of Blue InGaN/GaN Multi-Quantum Well Micro-LEDs. *Appl. Sci.* **2022**, *12*, 7431. [CrossRef]
6. Shi, J.; Ding, Y.; Tang, L.; Li, X.; Bai, H.; Li, X.; Fan, W.; Niu, P.; Fu, W.; Yang, X.; et al. Low-Frequency Terahertz Photonic Crystal Waveguide with a Lilac-Shaped Defect Based on Stereolithography 3D Printing. *Appl. Sci.* **2022**, *12*, 8333. [CrossRef]
7. Wang, J.; Zhang, W.; Wang, F.; Liu, Y.; Zhao, B.; Fang, X. Experimental Analysis and Verification of the Influence on the Elastic Recovery Coefficient of Wheat. *Appl. Sci.* **2023**, *13*, 5481. [CrossRef]
8. He, Y.; Yan, C.; Chen, K.; Xu, D.; Li, J.; Zhong, K.; Wang, Y.; Yao, R.; Yao, J.; Yao, J. High Repetition Rate, Tunable Mid-Infrared BaGa4Se7 Optical Parametric Oscillator Pumped by a 1 μm Nd:YAG Laser. *Appl. Sci.* **2022**, *12*, 7197. [CrossRef]
9. Wang, S.; Liu, Y.; Li, L. Sparse Weighting for Pyramid Pooling-Based SAR Image Target Recognition. *Appl. Sci.* **2022**, *12*, 3588. [CrossRef]
10. Shi, B.; Niu, J.; Zhou, X.; Dong, X. Quantitative Assessment Methods of Early Enamel Caries with Optical Coherence Tomography: A Review. *Appl. Sci.* **2022**, *12*, 8780. [CrossRef]

Article

Experimental Analysis and Verification of the Influence on the Elastic Recovery Coefficient of Wheat

Jizhong Wang, Weipeng Zhang, Fengzhu Wang, Yangchun Liu, Bo Zhao and Xianfa Fang *

National Key Laboratory of Agricultural Equipment Technology, China Academy of Agricultural Mechanization Science Group Co., Ltd., Beijing 100083, China
* Correspondence: fangboshi2023@163.com

Abstract: To establish a collision model of wheat grains impacting a force plate with a piezoelectric sensor, and to investigate the influence of the elastic recovery coefficient on the sensor's detection accuracy during the collision process, this study employed object kinematic principles to construct a wheat elastic recovery coefficient measurement device. This device ascertains the elastic properties of wheat during collisions and determines the elastic recovery coefficient of the wheat collision model. The wheat variety Jinan No. 17 was selected for testing, and the effects of the contact material, grain drop height, material thickness, and grain moisture content on the collision recovery coefficient during the collision process were analyzed through single-factor and multi-factor experiments. The experimental results demonstrate that the collision recovery coefficient of wheat grains increases with the stiffness of the collision materials for different materials. The grain recovery coefficient of wheat exhibits a downward trend with increasing falling height and moisture content, while it tends to rise as the material thickness increases. Data analysis and comparison reveal that, given the determination of the collision material, the moisture content of wheat exerts the most significant effect on the elastic recovery coefficient, followed by material thickness, while the influence of falling height is less pronounced. The findings of this study can provide data support for simulation testing and product design of wheat combine harvester cleaning screen body mechanisms and wheat seeders.

Keywords: response surface method; collision model; elastic recovery coefficient; elastic characteristics; kinematics

Citation: Wang, J.; Zhang, W.; Wang, F.; Liu, Y.; Zhao, B.; Fang, X. Experimental Analysis and Verification of the Influence on the Elastic Recovery Coefficient of Wheat. *Appl. Sci.* **2023**, *13*, 5481. https://doi.org/10.3390/app13095481

Academic Editors: Li Pei, Jia Shi, Hua Bai, Yunhui Mei and Pingjuan Niu

Received: 12 March 2023
Revised: 31 March 2023
Accepted: 4 April 2023
Published: 28 April 2023

1. Introduction

The elastic characteristics of wheat constitute fundamental data for designing cleaning shakers and sowing plates in wheat harvesters. During the screening operation of the harvester cleaning system, wheat undergoes collisions and ejections, either with other wheat particles or with the screen body, within the screening bin. The efficacy of the cleaning system's screen selection is influenced by these collision and ejection motions. Furthermore, an array of collision and ejection movements occurs as wheat is transported within the seed harvester, seed scraper, and seed feeder during seeding operations. Consequently, studying the elastic characteristics of wheat bears considerable significance for enhancing the vibratory screening mechanisms of wheat cleaning and the functional components of wheat seeders.

The recovery coefficient represents the capacity of an object to recover from deformation during collisions, first introduced by Newton as he employed the instantaneous impulse method to resolve the collision issue of rigid body systems. This coefficient serves as an essential parameter for characterizing alterations in the motion state of objects pre- and post-collision. Currently, extensive research has been conducted on elastic collision theory, the exploration recovery coefficient, and the methodologies for measuring these coefficients both domestically and internationally. Ning et al. assessed the elastic recovery coefficient for two soybean varieties [1]. Liu et al. analyzed and calibrated the elastic recovery coefficient for wheat seeds via a wheat accumulation test, subsequently deriving the discrete element parameters of wheat [2]. Li et al. measured maize seeds' recovery coefficient and established the contact parameters between maize seeds and seeders [3]. Kong investigated and analyzed the recovery coefficient for seed cotton; Yang et al. measured and analyzed the collision recovery coefficient for castor capsules [4]. Liu determined the collision recovery coefficient for oil sunflower grains [5]. Wen gauged the recovery coefficient of garlic seeds [6]. Zhang measured the recovery coefficient of mung beans [7].

In this study, Ji'nan 17 wheat seed was selected, and based on the analysis of the principle of kinematics, a wheat seed falling impact test platform was designed. The effects of the contact material, grain fall height, material thickness, and grain water content on the collision recovery coefficient of wheat were investigated through single-factor and multi-factor experiments, providing a basic data reference for the design optimization of harvesting and seeding machinery structure and the simulation modeling of loss sensor [8,9].

2. Materials and Methods

2.1. Test Materials

Four wheat samples with varying moisture content (10%, 13%, 15%, and 18%) were prepared by screening complete and plump wheat grains to simulate harvest conditions [10]. The working components of the combine harvester during wheat harvesting are typically composed of rubber, structural steel, and other materials. In order to simulate the collision recovery coefficient in the realistic environment of wheat grain collision, the test materials selected were structural steel Q235 and rubber. The material properties of the chosen materials can be ascertained by consulting the material library parameters as displayed in Table 1 [11].

Table 1. Collision material properties.

Material Model	Size mm (Length, Width, Thickness)	Density (g·cm^{-3})	Young's Modulus (MPa)	Poisson's Ratio
Q235 Steel	$140 \times 75 \times 7$	7.85	2.10×105	0.25
Rubber	$140 \times 75 \times 7$	1.8	100	0.30

2.2. Experimental Setup

To measure the collision recovery coefficient between wheat grain and collision material, a kinematic model of the grain impact process was established, and a test apparatus for grain elasticity recovery coefficient was constructed as illustrated in Figure 1. The test apparatus primarily consists of the overall support of the test bench, the grain collection plane, the collision plate or the loss sensor striking the force plate, the grain feeding mechanism, and the grain height lifting lead screw. The collision plate was mounted at a 45-degree angle in front of the grain collection plane of the test bench, and the wheat grain sample was positioned in the grain dispensing mechanism directly above.

2.3. Test Principle

Collision represents a prevalent mechanical phenomenon, characterized by a brief duration of the collision process [12], minimal displacement of the colliding objects, considerable velocity alterations, substantial impact force exerted by the colliding entities, and energy dissipation. Consider two objects possessing masses m_1 and m_2, colliding at velocities v_1 and v_2, respectively. In accordance with the conservation of momentum principle, the momentum conversion formula during the collision process involving these two objects can be deduced, as demonstrated in Equation (1) [13]:

$$\begin{cases} m_1v_1 + m_2v_2 = m_1v_1' + m_2v_2' \\ k = \frac{v_2' - v_1'}{v_1 - v_2} \end{cases} \tag{1}$$

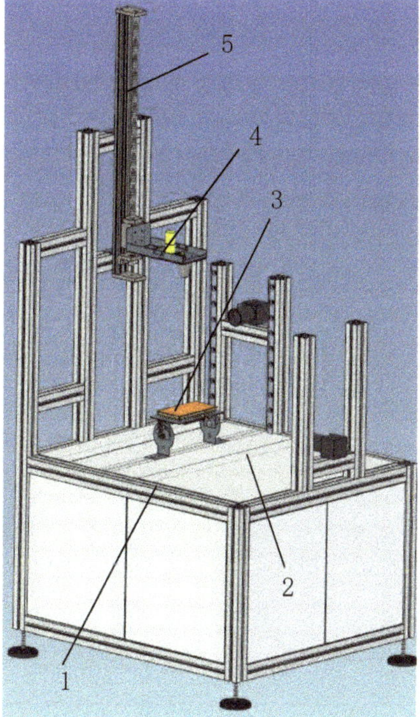

Figure 1. Grain impact test bench. 1. test bed bracket; 2. grain collection plane; 3. collision plate; 4. grain delivery mechanism; 5. grain height lifting screw.

After derivation and calculation, velocities v_1 and v_2 after the collision can be ascertained, as indicated in Equation (2):

$$\begin{cases} v'_1 = v_1 - (1+k)\frac{m_2}{m_1+m_2}(v_1 - v_2) \\ v'_2 = v_2 + (1+k)\frac{m_1}{m_1+m_2}(v_1 - v_2) \end{cases} \tag{2}$$

when k is equal to 1,

$$\begin{cases} v'_1 = v_1 - \frac{2m_2}{m_1+m_2}(v_1 - v_2) \\ v'_2 = v_2 + \frac{2m_1}{m_1+m_2}(v_1 - v_2) \end{cases} \tag{3}$$

when k is equal to 0,

$$v'_1 = v'_2 = \frac{m_1 v_1 + m_2 v_2}{m_1 + m_2} \tag{4}$$

where m_1 and m_2 denote the masses of the two colliding objects, with the unit in kg. v_1 and v_2 represent the pre-collision velocities of the two colliding objects, expressed in m/s; and v_1 and v_2 correspond to the post-collision velocities of the objects, with the unit in m/s. The reference formatting has been adjusted.

Upon analyzing Equations (3) and (4), it becomes apparent that the k value dictates the post-collision velocity alteration of the objects. When $k = 1$, a perfect elastic collision transpires, and the velocities of the two objects are transferred. Notably, when $m_1 = m_2$, the velocities of the two objects are exchanged following the collision. When $k = 0$, an imperfectly elastic collision occurs. Post-collision, the velocities become identical and the two objects proceed in unison.

In addition to their velocity, the kinetic energy of the two objects undergoes alteration following the collision, with the most pronounced change being kinetic energy loss.

Equation (5) illustrates the pre- and post-collision kinetic energy equation, with T_1 and T_2 representing the cumulative kinetic energy before and after the collision, respectively.

$$\begin{cases} T_1 = \frac{1}{2}m_1v_1^2 + \frac{1}{2}m_2v_2^2 \\ T_2 = \frac{1}{2}m_1v_1'^2 + \frac{1}{2}m_2v_2'^2 \end{cases} \tag{5}$$

The total kinetic energy loss of the two objects post-collision can be derived, as displayed in Equation (6):

$$\begin{aligned} \triangle T &= T_1 - T_2 \\ &= \frac{m_1m_2}{2(m_1+m_2)}(1+k)^2(v_1 - v_2)^2 \end{aligned} \tag{6}$$

when $k = 1$, kinetic energy loss is expressed as Equation (7):

$$\triangle T = T_1 - T_2 = 0 \tag{7}$$

where T_1 signifies the aggregate kinetic energy of the two objects prior to the collision, with the unit in J; T_2 represents the total kinetic energy of the two bodies following the collision, with the unit in J; and $\triangle T$ denotes the difference in kinetic energy between the two objects pre- and post-collision, expressed in J.

When $k = 0$, kinetic energy loss is expressed as Equation (8):

$$\triangle T = \frac{m_1m_2}{2(m_1 + m_2)}(v_1 - v_2)^2 \tag{8}$$

In accordance with the definition of the recovery coefficient e (both e and k are designated as elastic recovery coefficients), the proportion of the separation velocities of two objects in the normal direction at the contact point before and after collision represents the elastic recovery coefficient. Consequently, a schematic illustration of the determination of the elastic recovery coefficient is depicted in Figure 2. It is merely necessary to obtain the approaching velocity prior to the collision and the separating velocity following the collision to deduce the elastic recovery coefficient. To enhance the precision of velocity detection, this test employs the principle of kinematic equations, calculating the requisite velocity values through an indirect method.

After the grain experiences impact and rebound, a parabolic trajectory is formed, where s signifies the horizontal displacement following the grain rebound, and h represents the distance from the grain collection plane to the rebound point. From this, the x-axis directional velocity after the collision can be determined, as demonstrated in Equation (9):

$$v_x = s\sqrt{\frac{g}{2h}} \tag{9}$$

where s denotes the horizontal displacement of seeds post-rebound, with the unit being millimeters; h is the vertical distance between the rebound point and the grain collection plane (unit: mm). Additionally, g corresponds to neutral acceleration, measured in m/s²; v_x refers to the velocity along the x-axis after the grain collision, with the unit expressed in m/s.

The grain descends from the feeding mechanism, undergoing free-fall motion, with H being the distance from the seed falling point to the impact point. From this information, the y-axis directional velocity prior to grain collision can be calculated, as shown in Equation (10):

$$v_y = \sqrt{2gH} \tag{10}$$

where H is the height of grain fall and the unit is mm.

Finally, the formula for computing the recovery coefficient is derived by utilizing the definition of the recovery coefficient, as presented in Equation (11). In this equation, θ symbolizes the angle between the Y-axis velocity and the normal vector velocity before grain collision, while β represents the angle between the X-axis direction velocity and the normal vector velocity prior to grain collision.

$$e = \frac{v_1}{v_0} = \frac{v_x \cos \beta}{v_y \cos \theta} = \frac{s \cos \beta}{2 \cos \theta \sqrt{Hg}} \tag{11}$$

where e refers to the elastic recovery coefficient, θ is the angle between the velocity along the Y-axis and the normal vector preceding grain collision, and β is the angle between the velocity along the X-axis and the normal vector before the grain collision.

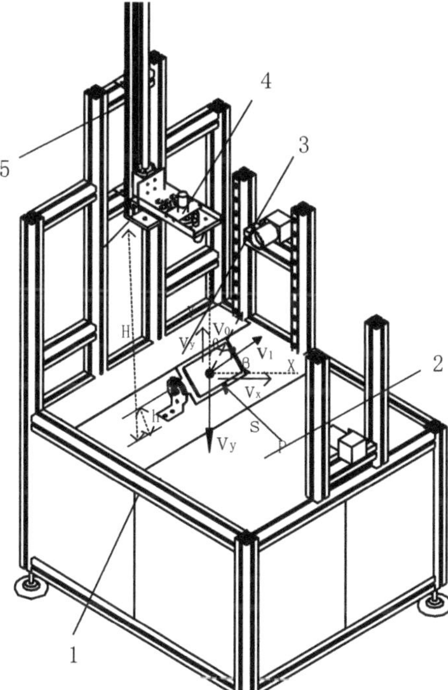

Figure 2. Schematic diagram of elastic recovery coefficient determination. 1. test bed bracket; 2. grain collection plane; 3. collision plate; 4. grain delivery mechanism; 5. grain height lifting screw.

2.4. Experimental Methods

The wheat recovery coefficient primarily correlates with moisture content, collision material, material thickness, and fall height of wheat. To emulate the actual harvest scenario, wheat samples with varying moisture contents of 10%, 13%, 15%, and 18% were employed in the experiment. The chosen materials comprise Q235 steel plate and rubber; the material thicknesses selected are 2 mm, 4 mm, and 8 mm; the seed drop heights are 120 mm, 180 mm, and 240 mm; and the factor levels constituting the wheat univariate test are presented in Table 2.

Table 2. The level of univariate test factors.

Trial Level	Water Content (%)	Collision Materials	Material Thickness (mm)	Drop Height (mm)
1	10	Q235 Steel	2	120
2	13	Rubber	4	180
3	15	/	8	240
4	18	/	/	/

/: indicates that no data exists.

In accordance with the factor level in Table 2, four sets of single-factor influence tests were conducted. In the test on the effect of moisture content on the recovery coefficient, Q235 steel served as the collision material, the material thickness remained at 4 mm, and the drop height was set at 180 mm, resulting in four sets of data. In the test regarding the impact of collision material on the wheat recovery coefficient, the material thickness remained at 4 mm, and the drop height was set at 180 mm, yielding eight sets of data. In the test regarding the effect of material thickness on the wheat recovery coefficient, Q235 steel served as the collision material, and the drop height was set at 180 mm, procuring 12 sets of data under distinct moisture conditions. In the test regarding the impact of fall height on the wheat recovery coefficient, Q235 steel served as the collision material, and the material thickness remained at 4 mm, obtaining 12 sets of data under varying moisture conditions.

3. Results and Analysis

3.1. Effect of Moisture Content on Recovery Coefficient of Wheat

The experimental data corresponding to various moisture content levels are presented in Table 3. From this information, a trend figure (Figure 3) illustrating the factors influencing single-factor water content on the recovery coefficient can be derived. The analysis reveals that under the test premise of employing Q235 steel as the collision material, maintaining material thickness at 4 mm, and setting the drop height at 180 mm, the recovery coefficient of wheat is significantly impacted by moisture content. As moisture content increases, the recovery coefficient progressively declines.

Table 3. Moisture content univariate experimental data.

NO	Whereabouts Height (mm)	Material Thickness (mm)	Collision Materials	Water Content (%)	Recovery Factor
1	180	4	Q235 Steel	10	0.4826
2	180	4	Q235 Steel	13	0.4651
3	180	4	Q235 Steel	15	0.4598
4	180	4	Q235 Steel	18	0.4276

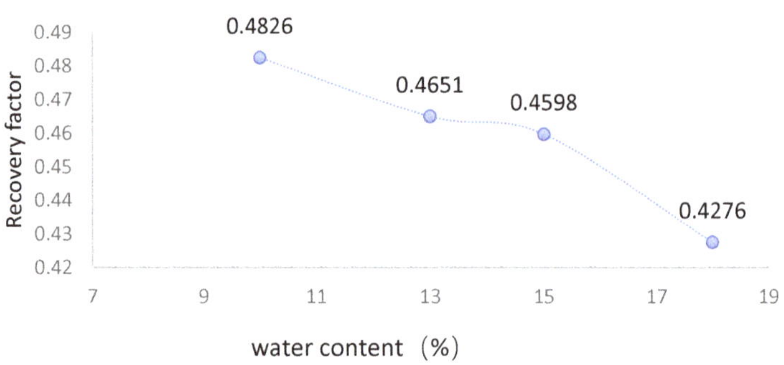

Figure 3. Trend of recovery coefficient of factors influencing single-factor moisture content.

3.2. Influence of Collision Materials on Wheat Recovery Coefficient

Table 4 displays experimental data related to different collision materials. A bar chart depicting the recovery coefficient of single-factor collision material can be generated (Figure 4). Analysis indicates that when material thickness remains fixed at 4 mm and the drop height is set at 180 mm for the test, the recovery coefficient of wheat is notably affected by the collision material. The recovery coefficient for grain collision involving rubber material across four moisture levels is lower than that of Q235 steel, which can be attributed to the soft texture of the rubber and the partial absorption of collision energy.

Table 4. Single-factor experimental data of collision materials.

NO	Whereabouts Height (mm)	Material Thickness (mm)	Water Content (%)	Collision Materials	Recovery Factor
1	180	4	10	Q235 Steel	0.4826
2	180	4	13	Q235 Steel	0.4651
3	180	4	15	Q235 Steel	0.4598
4	180	4	18	Q235 Steel	0.4276
5	180	4	10	Rubber	0.4716
6	180	4	13	Rubber	0.4521
7	180	4	15	Rubber	0.4306
8	180	4	18	Rubber	0.4086

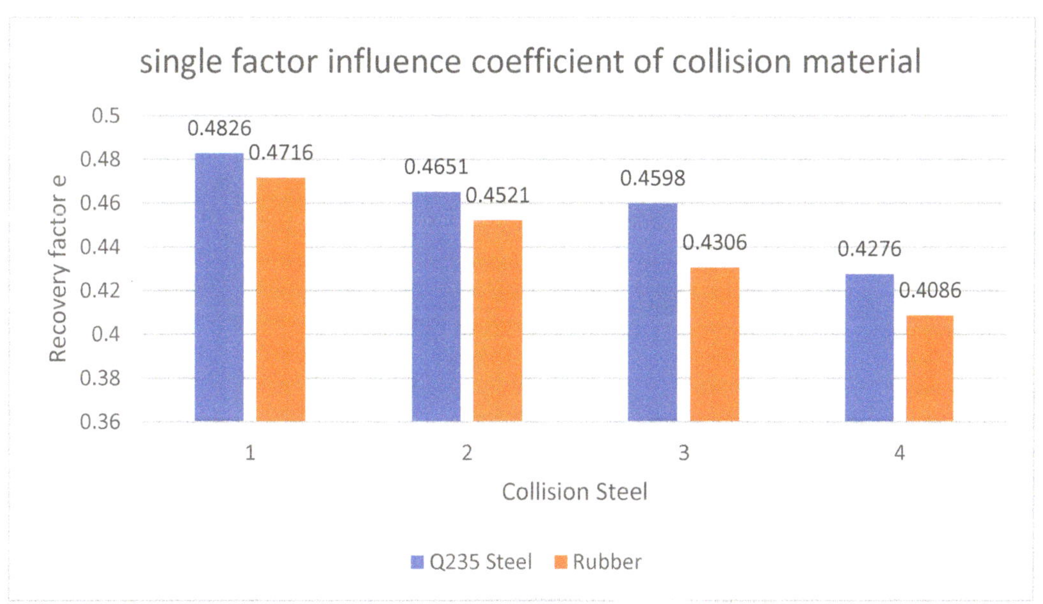

Figure 4. Single factor of collision material affects recovery factor.

3.3. Effect of Material Thickness on Wheat Recovery Coefficient

Table 5 presents experimental data for different material thicknesses. A distribution map of the factors influencing single-factor material thickness on the recovery coefficient can be created (Figure 5). Analysis demonstrates that under the test premise of utilizing Q235 steel for the collision material and maintaining the drop height at 180 mm, the recovery coefficient of wheat exhibits an increasing trend as material thickness grows. When the thickness reaches approximately 7 mm, the increase in recovery coefficient levels off and stabilizes. Material thickness directly reflects the stiffness properties of the material. As material thickness increases, stiffness follows suit. Consequently, when wheat deformation is reduced during collision events, energy loss also decreases, resulting in a larger elastic recovery coefficient.

Table 5. Material thickness unifactor experimental data.

NO	Whereabouts Height (mm)	Collision Materials	Material Thickness (mm)	Water Content (%)	Recovery Factor
1	180	Q235 Steel	2	10	0.4712
2	180	Q235 Steel	2	13	0.4535
3	180	Q235 Steel	2	15	0.4376
4	180	Q235 Steel	2	18	0.4109
5	180	Q235 Steel	4	10	0.4826
6	180	Q235 Steel	4	13	0.4651
7	180	Q235 Steel	4	15	0.4598
8	180	Q235 Steel	4	18	0.4276
9	180	Q235 Steel	8	10	0.4915
10	180	Q235 Steel	8	13	0.4721
11	180	Q235 Steel	8	15	0.4644
12	180	Q235 Steel	8	18	0.4309

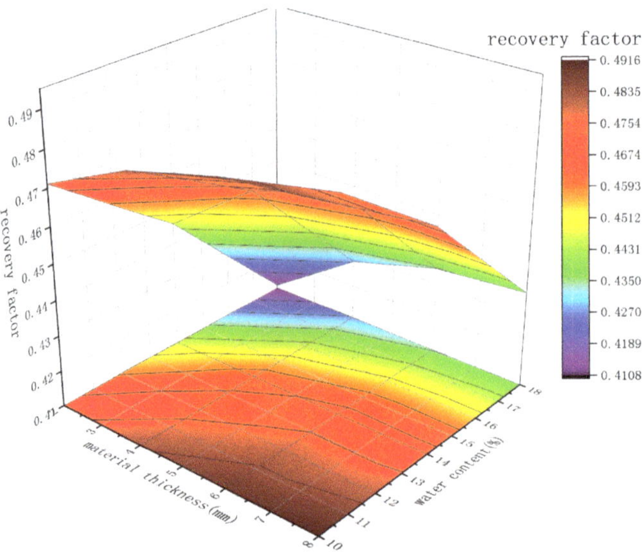

Figure 5. Single factor of material thickness affects the recovery factor.

3.4. Effect of Drop Height on Wheat Recovery Coefficient

The data from experiments with varying fall heights are presented in Table 6, from which a univariate fall height influence factor recovery coefficient distribution map can be derived, as depicted in Figure 6. Upon analysis, it becomes evident that, when utilizing Q235 steel as the collision material and maintaining a constant material thickness of 4 mm, the falling height of wheat grains exerts an influence on the recovery coefficient. For wheat with different moisture content, the impact of falling height on the recovery coefficient remains relatively consistent. As the fall height of wheat grains increases, the recovery coefficient between them and the collision material diminishes. This phenomenon can be attributed to the rising falling height of wheat particles, which leads to increased deformation during collisions with the test plate and heightened frictional resistance between the air and the collision plate. Consequently, the energy loss of wheat particles escalates during the collision process, causing the rebound velocity of grains to decrease after the collision and subsequently leading to a reduced recovery coefficient, as calculated based on the principle of elastic recovery coefficient.

Table 6. Single-factor experimental data of drop height.

NO	Collision Materials	Material Thickness (mm)	Drop Height (mm)	Water Content (%)	Recovery Factor
1	Q235 Steel	4	120	10	0.4968
2	Q235 Steel	4	180	13	0.4651
3	Q235 Steel	4	240	15	0.4368
4	Q235 Steel	4	120	18	0.4384
5	Q235 Steel	4	180	10	0.4826
6	Q235 Steel	4	240	13	0.4501
7	Q235 Steel	4	120	15	0.4735
8	Q235 Steel	4	180	18	0.4276
9	Q235 Steel	4	240	10	0.4685
10	Q235 Steel	4	120	13	0.4763
11	Q235 Steel	4	180	15	0.4598
12	Q235 Steel	4	240	18	0.4167

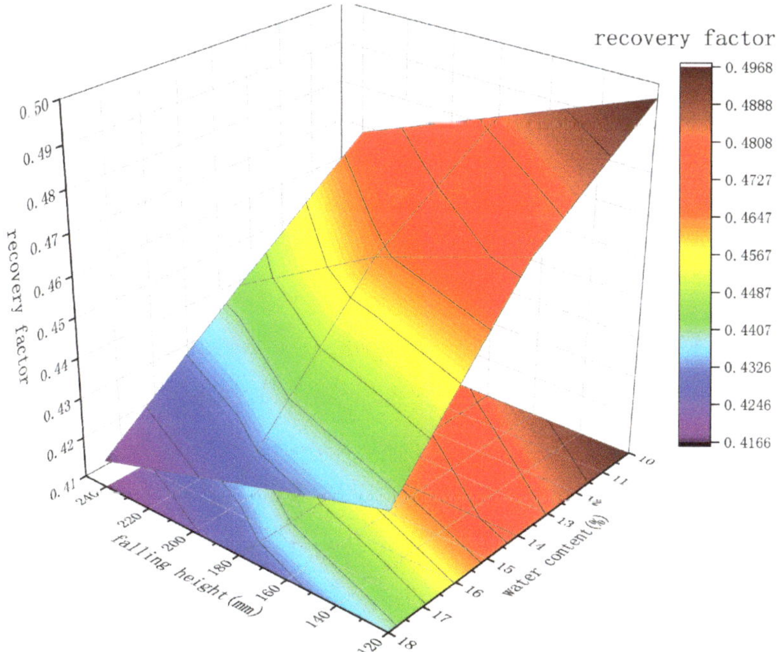

Figure 6. Single factor of fall height affects the recovery coefficient.

4. Conclusions

(1) The collision material influences the deformation and recovery capacity of wheat during collisions, serving as the primary factor affecting the recovery coefficient in wheat collision experiments. Given constant drop height, material thickness, and moisture content, the elastic recovery coefficient of wheat and rubber in this test is lower than that of wheat and Q235 steel. Therefore, during the mechanical design process for wheat, material selection should be tailored to the specific context.

(2) The falling height dictates the energy changes as the wheat grain collides with the test board material, particularly affecting the size of the wheat shape variable and the movement resistance. When the falling height increases, the grain shape variable expands, and the movement resistance rises correspondingly, leading to a reduced elastic recovery coefficient. The stiffness of the material is determined by its thickness, with an increase in material thickness enhancing the stiffness. When wheat grains come into contact with the

collision material, their energy loss decreases, the initial rebound velocity grows, and, in turn, the elastic recovery coefficient is elevated.

(3) Wheat grains interact with different materials during the collision process, undergoing force deformation and rebound recovery in two separate movement phases. Following the collision, wheat grains may also exhibit rotation and other factors. Consequently, the recovery coefficient of wheat for the same material with varying parameters in the collision material yields a wide range of recovery coefficient changes. The test process revealed that the same wheat grain produced varying test results in multiple tests.

Author Contributions: Conceptualization, J.W. and F.W.; methodology, Y.L.; software, W.Z.; validation, B.Z., X.F. and J.W.; formal analysis, W.Z.; investigation, Y.L.; resources, B.Z.; data curation, J.W.; writing—original draft preparation, J.W.; writing—review and editing, F.W.; visualization, Y.L.; supervision, W.Z.; project administration, J.W.; funding acquisition, W.Z. All authors have read and agreed to the published version of the manuscript.

Funding: This research was funded by the National Key Research and Development Program Project of China (2021YFD2000601).

Informed Consent Statement: Not applicable.

Data Availability Statement: Not applicable.

Conflicts of Interest: The authors declare no conflict of interest.

References

1. Ning, X.J.; Jin, C.Q.; Li, Q.L. Determination and analysis of physical characteristics of two soybean extractables in Huang-Huai-Hai area. *J. Agric. Mech. Res.* **2021**, *43*, 163–168, 175.
2. Liu, F.Y.; Zhang, J.; Li, B. Discrete element parameter analysis and calibration of wheat based on packing test. *Trans. Chin. Soc. Agric. Eng.* **2016**, *32*, 247–253.
3. Li, X.F.; Liu, F.; Zhao, M.Q. Determination of contact parameters between maize seeds and seeders. *J. Agric. Mech. Res.* **2018**, *40*, 149–153.
4. Kong, F.T.; Shi, L.; Zhang, Y.T. Determination and analysis of recovery coefficient in seed cotton collision model. *China Agric. Sci. Technol. Rev.* **2019**, *21*, 92–99. [CrossRef]
5. Liu, Y.; Zong, W.Y.; Ma, L.N. Determination of 3D collision recovery coefficient of oil sunflower seeds by high-speed photography. *Trans. Chin. Soc. Agric. Eng.* **2020**, *36*, 44–53.
6. Wen, E.Y.; Li, Y.H.; Niu, Z. Garlic seed particle discrete element model parameter calibration. *Agric. Mech. Res.* **2021**, *43*, 160–167. [CrossRef]
7. Zhang, Y.; Liu, F.; Zhao, M.Q. Experimental study on Physical and mechanical characteristics of Mung bean. *J. Agric. Mech. Res.* **2017**, *39*, 119–124.
8. Patil, D.; Higgs, C.F. Experimental investigations on the coefficient of restitution for Sphere-Thin plate elastoplatic impact. *J. Tribol.* **2018**, *140*, 011406. [CrossRef]
9. Chen, D.Y.; Yang, Z.; La, G.X. Determination of collision recovery coefficient in half-summer harvest. *Agric. Eng.* **2021**, *11*, 82–88. [CrossRef]
10. Zheng, H.B. *Simulation and Experimental Verification of Pyrolysis Motion-Heat Transfer Coupling Relationship in Waste Tire Rotary Kiln*; Southeast University: Nanjing, China, 2021.
11. Zheng, J. *Design and Test of American Ginseng Pneumatic Needle Precision Collector*; Huazhong Agricultural University: Wuhan, China, 2019.
12. Rezaiee-Pajand, M.; Mokhtari, M.; Masoodi, A.R. Stability and free vibration analysis of tapered sandwich columns with functionally graded core and flexible connections. *CEAS Aeronaut. J.* **2018**, *9*, 629–648. [CrossRef]
13. Li, Y.L.; Qiu, X.M.; Zhang, X. Analysis of different definitions of recovery coefficients and their applicability. *Mech. Pract.* **2015**, *37*, 773–777.

Article

Photothermal Determination of the Surface Treatment of Cd$_{1-x}$Be$_x$Te Mixed Crystals

Jacek Zakrzewski [1,*], Karol Strzałkowski [1,*], Mohammed Boumhamdi [1], Agnieszka Marasek [1], Ali Abouais [1,2] and Daniel M. Kamiński [3]

[1] Institute of Physics, Nicolaus Copernicus University, ul. Grudziadzka 5/7, 87-100 Torun, Poland
[2] Laboratory of Engineering Sciences for Energy, National School of Applied Sciences of El Jadida, BP 1166, El Jadida 24000, Morocco
[3] Institute of Chemical Sciences, Maria Curie-Skłodowska University, pl. Marii Curie-Sklodowskiej 3, 20-031 Lublin, Poland
* Correspondence: jzakrzew@umk.pl (J.Z.); skaroll@umk.pl (K.S.)

Abstract: Cd$_{1-x}$Be$_x$Te, a new material with potential for X-ray and γ-ray detectors, was analyzed by photothermal piezoelectric spectroscopy. The samples were tested depending on beryllium content and surface preparation. The main aim of the measurements was to extract the energy gap values, which were found for x = 0.01, 0.03, 0.05, and 0.1. It was shown that mechanical (polishing) and chemical (etching) treatment strongly influenced the amplitude and phase spectra of CdBeTe crystals. Piezoelectric spectroscopy allowed for comparing the quality of preparation of both surfaces for a single sample. The sub-surface damaged layer that was created as a result of surface processing had different thermal parameters than the bulk part of the sample. It was responsible for the additional peaks in the amplitude spectrum and changes in the phase spectrum of the photothermal signal. Two different methods of sample etching were analyzed. One completely quenched the signal, and the other did not eliminate the defects present on the surface after the cutting process. The article presents the preliminary interpretation of experimental data using the modified Blonskij model.

Keywords: photothermal spectroscopy; surface defects; semiconductors

Citation: Zakrzewski, J.; Strzałkowski, K.; Boumhamdi, M.; Marasek, A.; Abouais, A.; Kamiński, D.M. Photothermal Determination of the Surface Treatment of Cd$_{1-x}$Be$_x$Te Mixed Crystals. *Appl. Sci.* **2023**, *13*, 2113. https://doi.org/10.3390/app13042113

Academic Editors: Li Poi, Jia Shi, Hua Bai, Yunhui Mei and Pingjuan Niu

Received: 22 December 2022
Revised: 31 January 2023
Accepted: 2 February 2023
Published: 7 February 2023

1. Introduction

The quality of the crystals used in the construction of X-ray and gamma radiation detectors affects their efficiency and sensitivity; hence, there is a need to know about their lattice disorder, defects, and both radial and axial homogeneity. Good homogeneity and low defect density lead to significant charge transport properties, low leakage currents, and no conductive short circuits between the detector contacts. The substitution of the native element with a foreign atom within the crystal always leads to undesired effects like disordered structure, defect generation, etc.

Bulk form CdTe and Cd$_{1-x}$Zn$_x$Te mixed compounds are applied in X-ray and gamma-ray detectors [1–4], in electrooptic and photorefractive devices [5], and as substrates for epitaxy [6]. Cd$_{1-x}$Zn$_x$Te ($0.1 \leq x \leq 0.2$) crystals have been studied intensively over the last twenty years as materials for gamma-ray detection application and also spectroscopic X-ray imaging [7].

Intensive investigations of Cd$_{1-x}$Zn$_x$Te have also been carried out in recent years. Lately, C. Zhou et al., studied the extended defects in CdZnTe crystal [8], and the effects of Al-rich AlN transition layers on the performance of CdZnTe films for solar-blind photodetector were investigated by J. Gu [9]. Effects of deep-level traps on the transport properties of high-flux X-ray CdZnTe detectors were investigated by Y. Li et al. [10].

Great attention is also paid to the surface preparation of the samples [11–14]. Surface phenomena play a large role in obtaining crystals in all phases of processing and often affect the parameters of an electronic component. The sub-surface layer plays an important role, which in high-purity semiconductor devices could become significant due to miniaturization. The effects of the inductively coupled Ar plasma etching on the performance of (111) face CdZnTe detector were investigated by B. Song [15], who proposed and analyzed the influence of different solutions for etching. The proposed method could remove the damaged layer caused by mechanical polishing. Still, it also led to the surface composition deviating from the stoichiometric ratio and forming a Te-rich surface. They noticed that the Te-rich layer is a highly conductive region, which results in a large surface leakage current and affects the detection performance. Zhang et al. [16] proposed a new chemical mechanical polishing method (silica, hydrogen peroxide, and citric acid), effectively reducing surface roughness. Min et al. [17] analyzed the effect of hydrogen plasma on CdZnTe and showed that hydrogen plasma could fill the Cd vacancy defect on the surface and reduce the leakage current.

Photothermal spectroscopy was also used to investigate the surface treatment procedures and their influence on the surface quality. Zakrzewski et al. [18] applied piezoelectric phase spectra to determine both the energy gaps (E_g) and the thermal diffusivities of $Cd_xZn_{1-x}Se$ mixed crystals. The authors showed and interpreted the different amplitude and phase spectra characters depending on the surface preparation procedure for $Cd_{0.3}Zn_{0.7}Se$ and $Cd_{0.5}Zn_{0.5}Se$ samples. The mechanical (grounding, polishing) and chemical (etching) procedures of surface preparation and their impact on both the amplitude and phase of photothermal spectra were observed for $Zn_{1-x-y}Be_xMn_ySe$ compounds [19]. The surface defects in the samples' sub-surface damaged layer were also reported for ZnSe binary crystal [20].

There are many methods of surface quality testing, e.g., different imaging methods. Piezoelectric photothermal spectroscopy is a cheap and fast method. Other photothermal methods that use microphones or pyroelectrics as detectors are not sensitive (do not show) to effects related to surface states. This is an advantage and a disadvantage; in the case of thermal diffusivity studies in the frequency domain, there is no signal coming from the sub-surface layer, and it is easier to interpret the obtained results (there are well-known models).

Still, more investigations and literature data on $Cd_{1-x}Be_xTe$ mixed crystals are needed. One article has been published so far [21], which determines thermal properties using the pyroelectric method in the frequency domain. The present work investigates the optical and thermal properties using piezoelectric photothermal spectroscopy. Adding beryllium to the CdTe matrix should give a similar effect in tuning the energy bandgap or lattice constant as in the case of zinc. This fact should be interesting for detecting purposes and producing better substrates for infrared sensors.

Here we report the first experimental data describing the properties of crystals grown in our laboratory. Piezoelectric photothermal spectroscopy was applied to characterize $Cd_{1-x}Be_xTe$ mixed crystals. This work aimed to investigate the optical properties of $Cd_{1-x}Be_xTe$ mixed crystals and the effect of surface treatment on the amplitude and phase spectra. On that basis, we aimed to find a procedure and the best solution for obtaining excellent surface quality among the samples.

2. Materials and Methods

Crystals investigated in this work were grown by using the vertical Bridgman technique. Before the growth process, starting mixtures were prepared from high-purity CdTe (6 N) and Be (99.9%) powders and put into a graphite crucible. The crucible was placed into a growth chamber, sealed, and evacuated. After evacuation, the external pressure of argon gas (around 90 atm.) was applied to reduce evaporation of the material during the growing. The temperature in the hot zone was set to 1600 K, with a stability better than 0.1 K. We used a 4 mm/h pulling rate to grow the crystal within two days. Typically our crystals grown by the Bridgman high-temperature and high-pressure technique were 4–6 cm in length and 1 cm in diameter. We could cut 15–20 plates 1–1.5 mm thick from one crystal rod. Four specimens of beryllium content (in ingot) were measured for x = 0.01, 0.03

0.05, and 0.1. Pure CdTe crystals were also measured for the same procedure of surface preparation. More details concerning the growth procedure can be found elsewhere [22].

The beryllium concentration in the samples was determined after crystal milling (in agate mortar) by applying the powder X-ray diffraction technique (Empyrean diffractometer, Malvern Panalytical) with a Cu anode used as a source of CuKα X-ray radiation (λ = 1.5406 Å). More details about the measurement were presented in previous work [23]. The diffraction data were fitted using the ReX v. 0.91 Rietveld analysis software. The following parameters were fitted: scale factor, 2-theta offset, background coefficients, and cell constant a for the CdTe structure. In this case, the beryllium in the CdTe was treated as a solid solution of the first type, and thus the main changes were mainly observed for the cell dimension. All U factors were isotropic and set to 0.5. The reflection shape was described by the pseudo-Voight function. For analysis, the necessary crystal structure file of CdTe (cif) was downloaded from the American Mineralogist Crystal Structure Database [24].

The XRD patterns of example CdBeTe mixed crystals (CdTe and $Cd_{0.9}Be_{0.1}Te$) powdered are presented in Figure 1a. Characteristic diffraction reflections from (111), (220), (311), (400), (331), (420), and (511) planes of the cubic phase (type of zinc blende structure) are visible on the patterns for both CdTe (black) and $Cd_{0.9}Be_{0.1}Te$ (red) samples. In the case of a mixed crystal, the diffraction peaks are shifted towards higher 2 θ in comparison to the position of these signals in the CdTe. Such behavior demonstrates the presence of mixed crystals, so the beryllium was successfully incorporated into the crystal lattice of CdTe. The absolute values of the lattice constant were calculated by the Rietveld method and the results that were obtained as the function of the composition are displayed in Figure 1b. Linear dependence of the lattice constant versus Be content confirms the composition of the grown crystals.

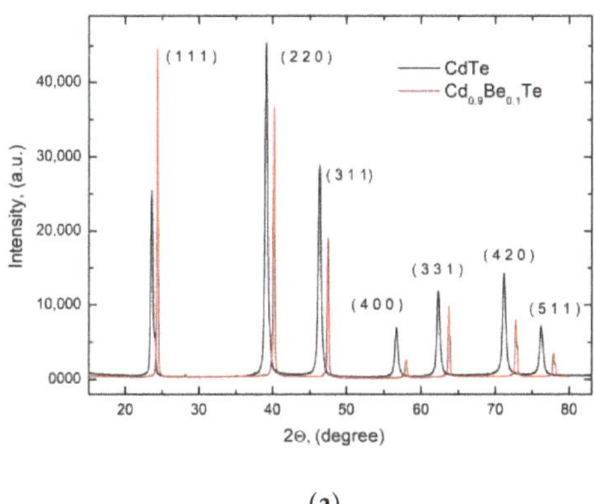
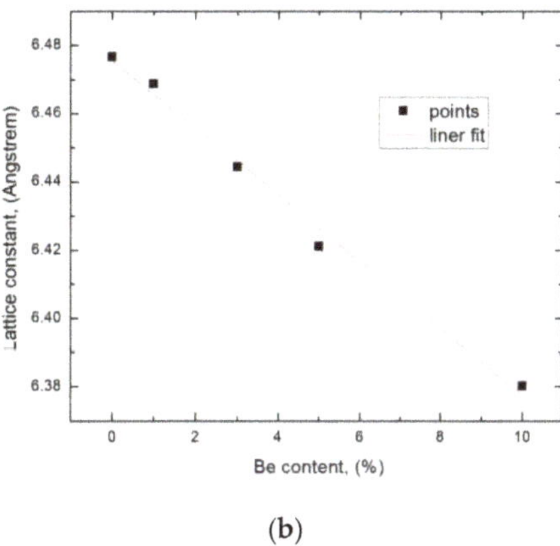

(a)

(b)

Figure 1. The XRD patterns of CdTe (black) and $Cd_{0.9}Be_{0.1}Te$ (red) powders (**a**) and lattice constant versus beryllium concentration (**b**).

The cut samples were ground in a suspension of water and Al_2O_3 powder on flat glass parallel plates. The grain diameter of the grinding powder was approximately 10 μm. Then, the samples were polished with diamond paste on polishing discs until a mirror surface was obtained. The grain diameter in the polishing paste was between 0.1 and 1 μm.

The samples were subjected to two different etching processes. First, the samples were etched using a mixture containing 48% HF (hydrofluoric acid), 30% H_2O_2, and H_2O. Digestion took place at room temperature. The samples were subjected to photothermal tests. Then, the surfaces of the samples were ground and polished again to use a mixture of $K_2Cr_2O_7$, 96% H_2SO_4, and H_2O for etching. The etching took place at a temperature of about 80 °C. After etching, the samples were briefly immersed in a 50% NaOH solution, then rinsed in water, and finally in ethyl alcohol.

The typical experiment setups for piezoelectric detection, piezoelectric cell allowing front and rear detection, and lock-in detection were used [25]. A 150 W xenon lamp was used as the light source. The beam passed through a computer-controlled spectrometer; 90% of the beam fell on the photothermal cell and 10% on the photodiode, whose task was to monitor the light intensity. Cell and photodiode signals were measured by two lock-in Stanford SR 510 units. In piezoelectric detection, there are two possible ways of mutual arrangement of the detector and the measured sample. The detector is located behind the illuminated sample in the more commonly used rear configuration. In the front configuration, the detector is placed on the illuminated surface. Depending on the configuration, different course amplitude and phase spectra are observed.

3. Results

3.1. Cut and Grounded Samples

First, cut and grounded samples were examined. Figure 2 presents the experimental amplitude (blue) and phase (green) spectra of $Cd_{0.97}Be_{0.03}Te$ in the rear configuration at the frequency modulation of 126 Hz. In both spectra, the theoretical simulations for the ideal sample are presented (in black). The different characters of both experimental and simulated spectra are visible. There are two additional maxima in the sub-bang gap region for the amplitude and the significant change (at the energy of E = 1.46 eV) of the phase in the same area. The amplitude signal in the above bandgap area is increasing in comparing stable values in simulated amplitude spectra. Similar behavior was previously observed for $Zn_{1-x-y}Be_xMn_ySe$ compounds [19]. It was associated and interpreted as coming from the defects located on the surface of the sample and strictly connected with the method of surface preparation.

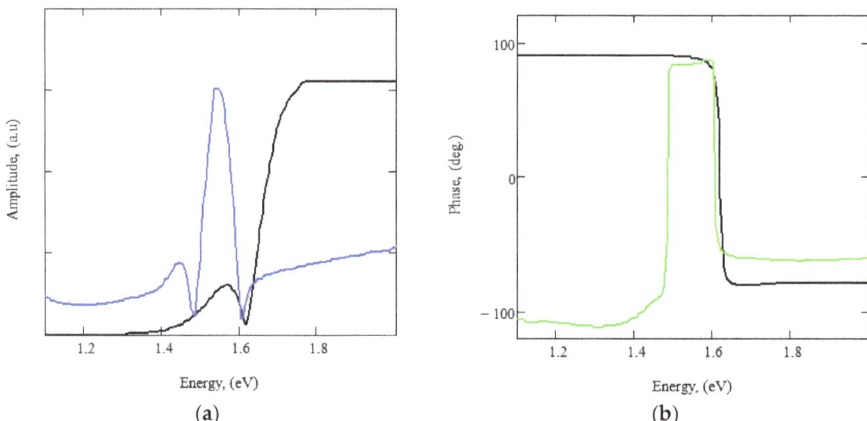

(a)

(b)

Figure 2. (**a**) Experimental amplitude (blue) and (**b**) phase (green) spectra of $Cd_{0.95}Be_{0.05}Te$ in the rear configuration at the frequency modulation of 126 Hz. The solid black lines simulate the ideal crystal's theoretical amplitude and phase spectra.

Figure 3 presents amplitude (a, b) and phase (c, d) spectra for $Cd_{0.99}Be_{0.01}Te$ grounded mixed crystals for different frequencies of modulation frequency and the rear (a, c) and front (b, d) configurations. In this case, the character of amplitude and spectra deviates from the theoretical predictions. For the lowest frequency (12 Hz), the additional maximum visible for higher frequencies is not visible. The higher the frequency, the higher the intensity of the maximum below the energy gap will be compared to the area above the energy gap: as the modulation frequency increases, the thermal diffusion length decreases, and the signal is generated from a smaller thickness. For low frequencies, the signal comes from a large sample thickness, and the signal from the surface is not dominant. It is worth noting that regardless of the frequency, the points of inflection of the curves are observed in the phase spectra for the same energies.

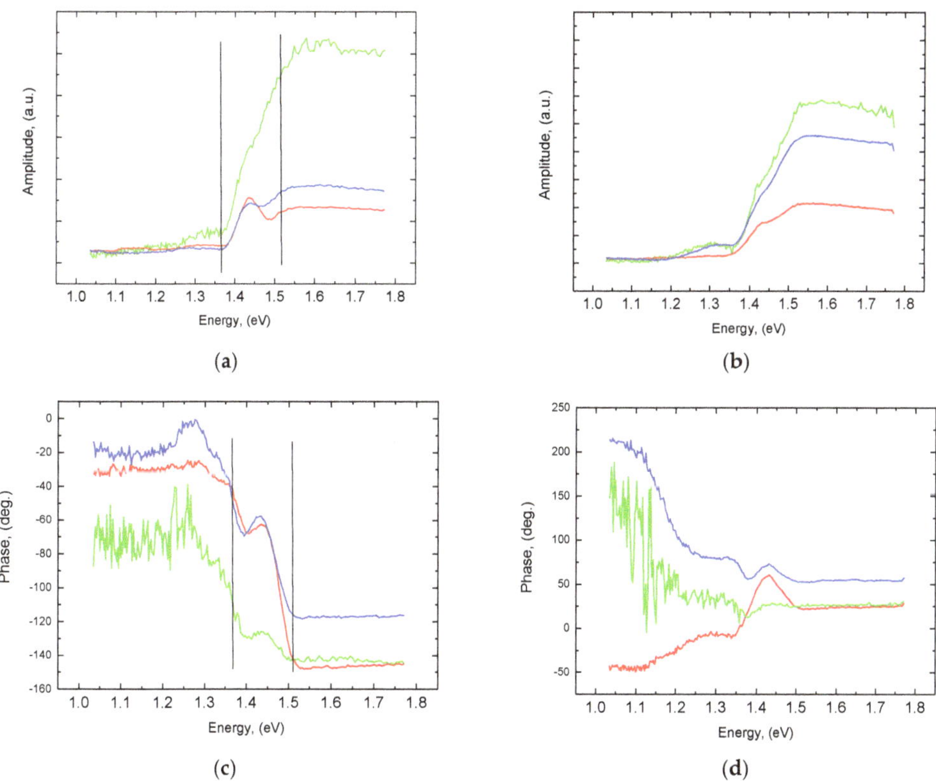

Figure 3. Amplitude (**a**,**b**) and phase (**c**,**d**) spectra for $Cd_{0.99}Be_{0.01}Te$ grounded mixed crystal for different frequencies of modulation frequency for the rear (**a**,**c**) and front (**b**,**d**) configurations at 12 Hz (blue lines), 76 Hz (green lines), and 126 Hz (red lines).

The different amplitude and phase spectra characters were observed for the $Cd_{0.9}Be_{0.1}Te$ mixed crystal. Figure 4 presents amplitude (a) and phase (b) spectra for this sample.

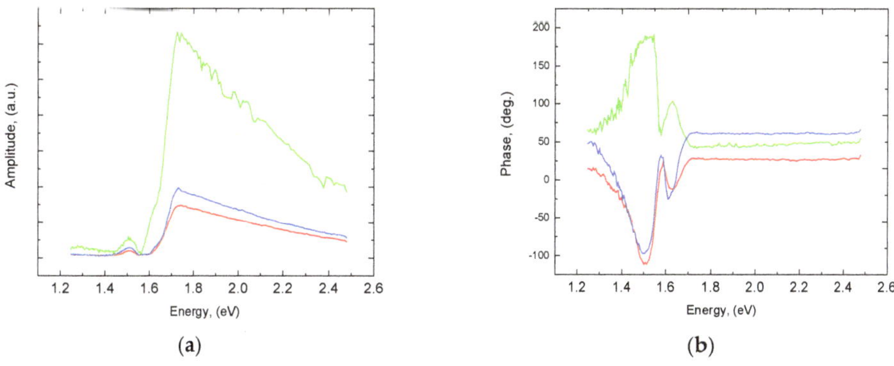

Figure 4. Amplitude (**a**) and phase (**b**) spectra for $Cd_{0.9}Be_{0.1}Te$ grounded mixed crystal for different frequencies of modulation frequency for the front configuration at 12 Hz (blue lines), 36 Hz (cyan line), 76 Hz (green lines), and 126 Hz (red lines).

Two maxima are observed for this sample in the amplitude spectra of $E_1 = 1.51$ eV and $E_2 = 1.59$ eV. The first one is present for all the frequencies, and its intensity is not changed compared to the intensity of the signal in the region above the energy gap. The behavior of the second maximum is similar to the one observed in the previous sample. A probable cause is that the thickness of the damaged layers on both surfaces is different. In the phase spectra, one must analyze the maxima and bending points to identify the localization of defects and their energetic positions.

Beryllium and defects related to its presence in the subsurface layer are undoubtedly responsible for changes in spectra and the formation of additional peaks. A strong correlation between surface preparation for $Zn_{1-x-y}Be_xMg_ySe$ [26] was previously observed, but no explanation has yet been found for the mechanisms of the formation of defects on the surface; research is ongoing to clarify their nature.

A decrease in amplitude spectra in the region above E_g (not a constant signal as in theory) was observed by Kuwahata et al. [27] in studies of implanted Si + ions in InP. According to them, silicon ions are responsible for defects in the crystal lattice and suppress the propagation of an elastic wave, which causes a decrease in the amplitude above the energy gap region. This is related to an increase in the number of free carriers in the conduction band states in the annealed samples. When these states are filled, photons of the energy close to the energy gap are not absorbed, and the photothermal signal is not generated.

A similar effect was also reported by Matsumori et al. [28] in studies of the impact of damage caused by ion implantation in Si. The photothermal signal was sensitive to the size and structure of damage in the layer with implanted ions. The authors attributed the signal reduction below the energy gap of the surface quality and the defects present on it. Its damage suppresses the generation of elastic waves. The authors came to conclusions that can also be drawn from the current research: the increased number of defects on the surface causes a more significant generation of thermal energy in the sub-bandgap region and the opposite effect for the area above the gap. Here, the thermoelastic waves are suppressed.

3.2. The Change of the Energy Gap with the Be Content

The series of $Cd_{1-x}Be_xTe$ samples was measured to obtain the energy gap values' dependence on the beryllium content described by the composition parameter (x). The amplitude and phase spectra were measured for the samples: $x = 0.01, 0.03, 0.5, 0.1$. A pure CdTe sample was also measured and subjected to the same surface treatment procedure. The amplitude and phase spectra for the modulation frequency of 126 Hz are presented in Figure 5. There was an increase in the value of E_g with the rise of beryllium content (clearly visible). Calculating the energy gap value from the phase and amplitude spectra was possible. The values could be estimated directly from the spectra or calculated as a parameter from the theory.

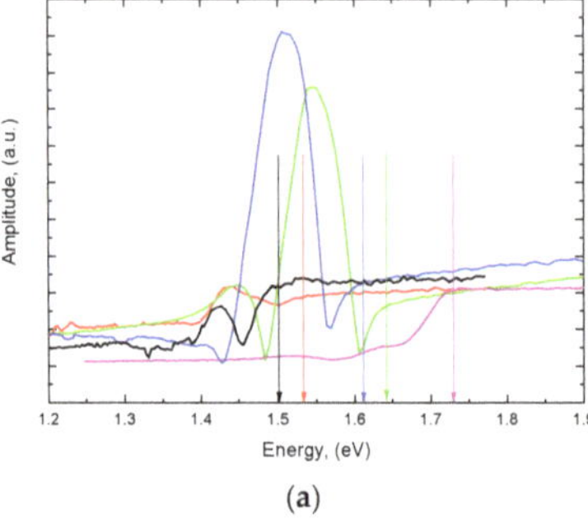

(a)

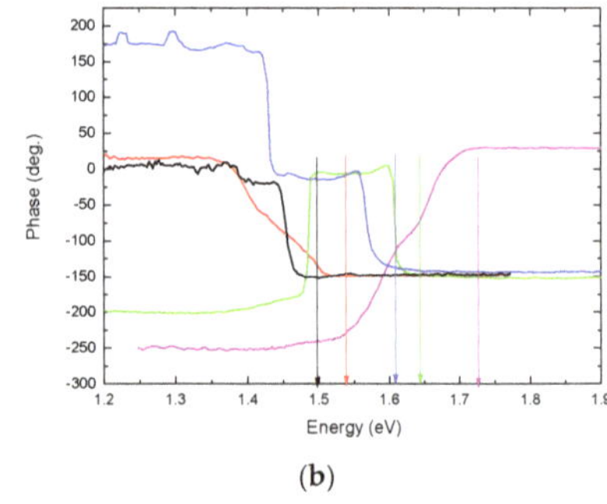

(b)

Figure 5. Amplitude (**a**) and phase (**b**) spectra of $Cd_{1-x}Be_xTe$ for different beryllium content. Black line: x = 0, red: x = 0.01, blue: x = 0.03, green: x = 0.05, magenta: x = 0.1. The color arrows indicate the values of energy gaps.

The phase spectra course may differ for different samples and frequencies. Still, the most important are the changes (inflection points) at which these changes occur: they indicate additional photothermal signal sources and the energetic position of the defects on the surface.

The obtained values of energy gaps are presented in Table 1.

Table 1. The experimental values of E_g for $Cd_{1-x}Be_xTe$.

x	Energy Gap (E_g), eV
0	1.49 eV
0.01	1.53 eV
0.03	1.61 eV
0.05	1.64 eV
0.1	1.72 eV

3.3. Etched Samples

One of the sample preparation goals was to obtain a perfect-quality surface. First, the polished samples were subjected to etching using a mixture containing 48% HF (hydrofluoric acid), 30% H_2O_2, and H_2O. This procedure caused the photothermal signal to be wholly quenched for mixed samples, making it impossible to measure the spectra. A possible explanation for this will be discussed later in this article. Therefore, it was decided to grind the samples again and further etch them in a new solution. The surfaces of the specimens were ground and polished to use a mixture of $K_2Cr_2O_7$, 96% H_2SO_4, and H_2O for the etching.

Figure 6 compares the amplitude spectra for the ground and etched $Cd_{0.95}Be_{0.05}Te$ sample.

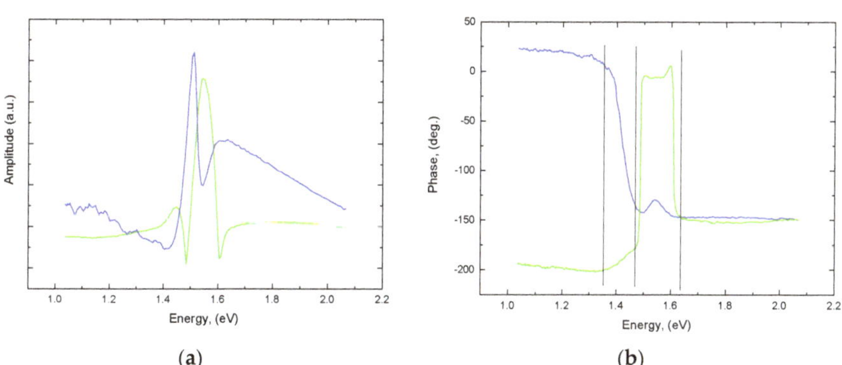

(a)

(b)

Figure 6. Amplitude (**a**) and phase (**b**) spectra for the ground (green lines) and etched (blue lines) $Cd_{0.95}Be_{0.05}Te$ sample at 126 Hz of modulation frequency.

The expectation was to obtain the spectra close in shape to the ideal crystal (black in Figure 2). The applied method of etching did not improve the surfaces of the sample. In the amplitude spectrum, an additional maximum still exists below the energy gap, which is not expected for an ideal sample. Although the phase has a different shape, it is worth noting that phase changes take place at the same energy values (indicated in the figure with vertical lines). This proves that the defects on the surface were present at a different depth of the damaged layer (by etching) or that additional effects had been achieved, e.g., with phenomena associated with the carriers. The same changes in phase are manifested differently in amplitude; with different maximums, it is also worth emphasizing that piezoelectric spectroscopy allows for studying the differences on both surfaces of the sample prepared with the same treatment procedure.

Figures 7 and 8 present the experimental spectra of the amplitude and phase of $Cd_{0.95}Be_{0.05}Te$ after the etching process was obtained for the illumination of different surfaces of the sample. The sample was measured in rear and front configurations. The surfaces of the samples were assigned numbers 1 and 2 and then measured after illumination of surface 1, then 2. The aim was to examine and compare the surfaces after an identical treatment procedure.

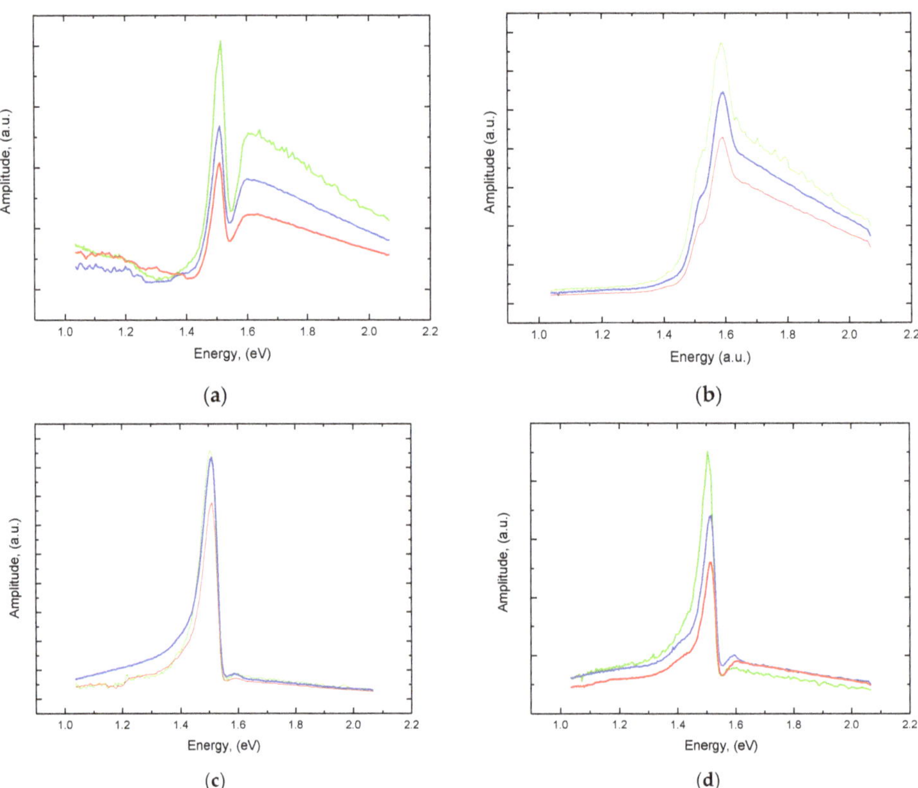

Figure 7. Amplitude spectra of $Cd_{0.95}Be_{0.05}Te$ for front and rear detection and illumination of different surfaces at 12 Hz (blue lines), 76 Hz (green lines), and 126 Hz (red lines). (**a**) Surface 1 illuminated, rear detection. (**b**) Surface 1 illuminated, front detection. (**c**) Surface 2 illuminated, rear detection. (**d**) Surface 2 illuminated, front detection.

Complementarity, especially in the amplitude spectra, is evident. Whether rear or front detection is considered, the strong maximum associated with the surface defect is observed for the illumination at surface 2. Its intensity is much higher in comparison to the above bandgap region, which indicates that etching negatively affects the quality of the surfaces, and the defects strengthen the photothermal signal. In the case of illumination of surface 1, the amplitude spectra have a similar character with a smaller maximum in the sub-bandgap region. The maximum has a higher intensity for rear detection and is associated with subtracting the piston and drum effects.

The phase spectra for the illumination of surface 2 show the changes at the same energy values despite the curves' different characters. The more significant differences are observed for phase spectra for front detection and the illumination at surface 1. It should be emphasized that both phase and amplitude must be analyzed to interpret piezoelectric spectra correctly.

The same change of character was also observed for the rest of the samples, which proved that the chosen etching procedure introduced additional defects on the surfaces of the samples.

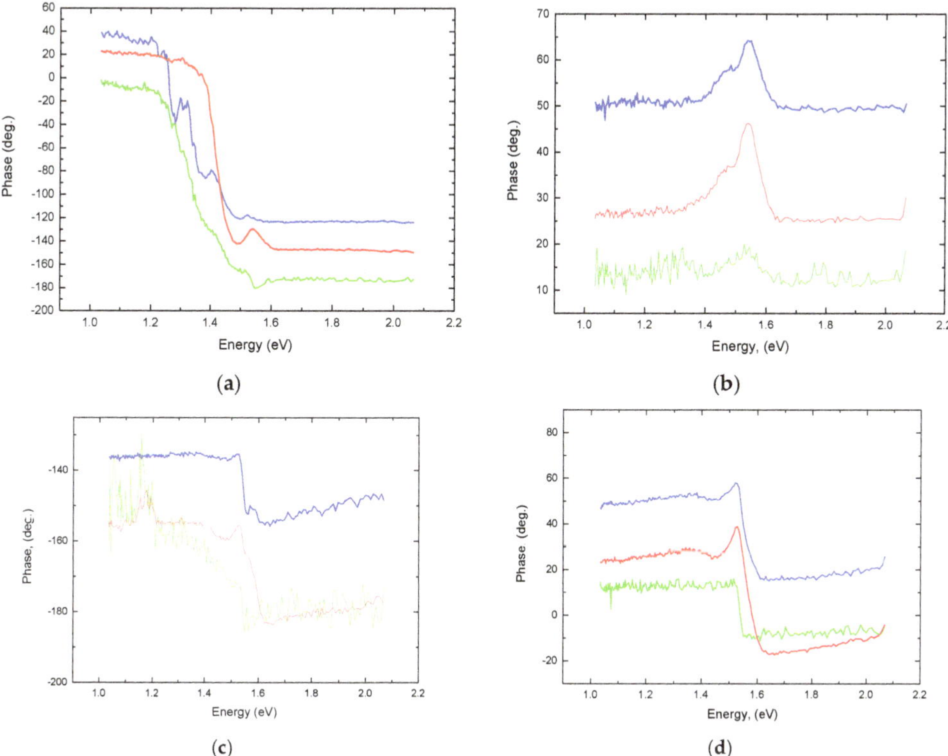

Figure 8. Phase spectra of Cd$_{0.95}$Be$_{0.05}$Te for front and rear detection and illumination of different surfaces at 12 Hz (blue lines), 76 Hz (green lines), and 126 Hz (red lines). (**a**) Surface 1 illuminated, rear detection. (**b**) Surface 1 illuminated, front detection. (**c**) Surface 2 illuminated, rear detection. (**d**) Surface 2 illuminated, front detection.

4. Discussion

The amplitude and phase spectra courses for the ideal semiconductor are known [25]. To measure against them, the temperature distribution in the tested sample and the waveforms for the absorption coefficient below and above the band gap were required. The expression for the absorption coefficient is one of the parameters of the temperature distribution. The expression for the phase and amplitude of the photothermal signal depends on this distribution. Details are presented in prior work [25].

The changes observed in the experimental spectra were interpreted as originating from defects located on the surface of the samples, as they were observed to be strongly dependent on the method of surface preparation. To simulate the changes, it was assumed that the defects could be simulated in temperature distributions with Gauss-shaped expressions. In the simulations, parameters such as energy, thermal diffusivity, and thermal conductivity were determined, and their courses were compared with the obtained experimental spectra.

The modified Blonskij's model [26,29,30] was applied to interpret the obtained spectra. The heat conduction equation describes the sample's temperature distribution as previously described [30]. The modified model considers a damaged subsurface layer on both surfaces of the sample with significantly different thermal parameters from the main part of the sample. The absorption coefficient due to the presence of the defects has the Gaussian character.

$$\beta = A_d \cdot exp\left(\frac{E - E_d}{\beta_1}\right)^2 \tag{1}$$

where E_d is the value of the energy of the defect, β_1 is the parameter describing the width of Gaussian shape maximum, and A_d is the amplitude of the maximum.

The defect is located at the surfaces of the sample, and its nature can be associated with the quality of the surface after the preparation process (grounding, polishing, etching). For the energy E_d of radiation, the layer strongly absorbs, while the volume of the sample is transparent. The temperature distribution in the sample is the sum of temperatures generated on the surface and the volume of the sample:

$$T'(x) = T(x) + T_1^d(x) + T_2^d \tag{2}$$

The piezoelectric signal is given by the expression [24]

$$V \sim \left(\frac{1}{l} \int_{-l/2}^{l/2} T'(x)dx \pm \frac{6}{l^2} \int_{-l/2}^{l/2} T'(x)xdx \right) \tag{3}$$

Two terms in the expression represent piston and drum effects [26]; they add up in the case of front detection and subtract in the case of rear mode.

Assuming the proper expressions for the absorption coefficient in the direct bandgap semiconductors (Urbach tail thermal broadening and absorption connected with the band-to-band electron transitions) [26], one can simulate the amplitude and phase spectra for the semiconducting sample with the presence of the defects localized on the subsurface damaged layer.

The nature of the amplitude and phase spectra is influenced by thermal bulk and layer parameters, the most important of which are thermal diffusivities, energy gaps, and the thicknesses of the sample and the near-surface layer. These basic parameters characterizing the material were used to simulate both amplitude and phase spectra.

Figure 9 presents amplitude (a) and phase (b) spectra of ground $Cd_{0.95}Be_{0.05}Te$ mixed crystal. Experimental data are shown together with simulations according to the above-given model. The simulation was performed for the presence of two defects with the location E_{d1} = 1.51 eV and E_{d2} = 1.55 eV, thermal diffusivity of 0.05 cm^2/s, thermal conductivity of 0.07 W/(cm·K) for the sample, and conductivity of the layer, ten times smaller than the bulk: A_{d1} = 15 cm^{-1}, A_{d2} = 7 cm^{-1}, β_1 = 0.2 cm^{-1}, β_2 = 0.06 cm^{-1}. The thickness of the defective layer was 0.01 mm. The simulations qualitatively confirm the presence of defects on the surface, but they are not entirely consistent with the experiment data.

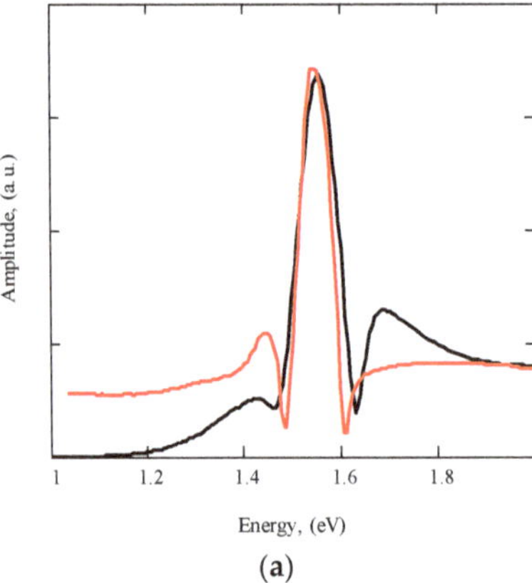

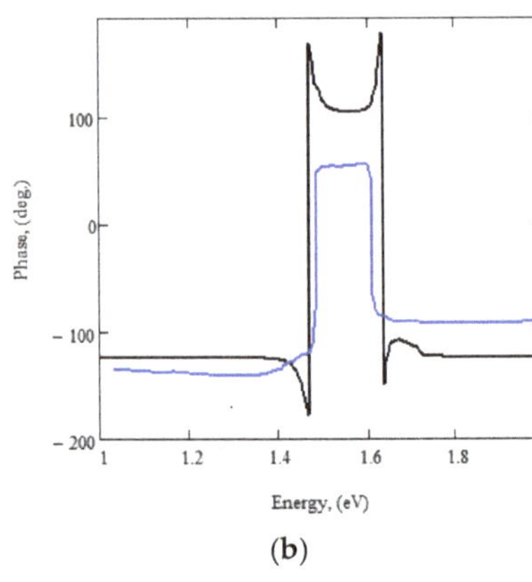

Figure 9. Amplitude (**a**) and phase (**b**) spectra of the ground $Cd_{0.95}Be_{0.05}Te$ mixed crystal for 126 Hz modulation frequency. Simulated spectra are presented by black lines, experimental amplitude by red lines, and experimental phase by blue lines.

Figure 10 presents amplitude (a) and phase (b) spectra of etched $Cd_{0.95}Be_{0.05}Te$ mixed crystal. Experimental data are shown together with simulations. The simulation was performed for the presence of two defects with a location different than the previous case: $E_{d1} = 1.6$ eV and $E_{d2} = 1.62$ eV, thermal diffusivity of 0.05 cm^2/s, thermal conductivity of 0.07 W/(cm·K) for the sample, and conductivity of the layer, 12 times smaller than the bulk: $A_{d1} = 110$ cm^{-1}, $A_{d2} = 80$ cm^{-1}, $\beta_1 = 0.045$ cm^{-1}, $\beta_2 = 0.06$ cm^{-1}. The thickness of the defective layer was 0.001 mm, ten times smaller than in the previous case. In this case, lower compatibility in experiment and simulation was achieved.

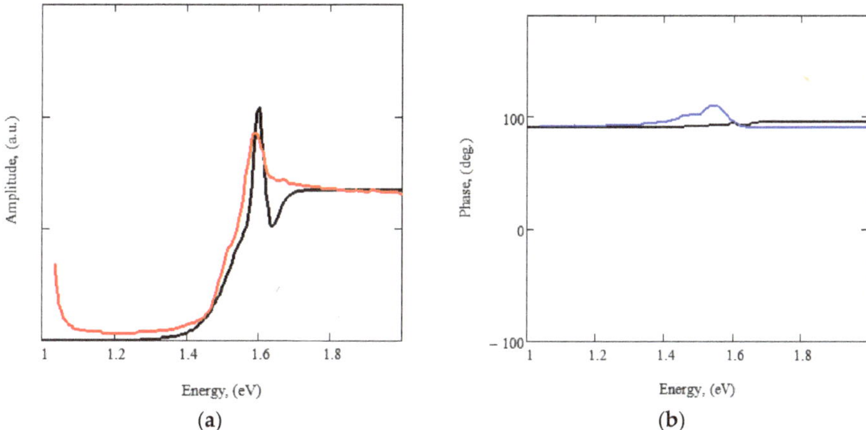

Figure 10. Amplitude (**a**) and phase (**b**) spectra of the etched $Cd_{0.95}Be_{0.05}Te$ mixed crystal for 126 Hz modulation frequency. Simulated spectra are presented by black lines, experimental amplitude by red lines, and experimental phase by blue lines.

The applied theory assumes the generation of a photothermal signal only due to the thermoelastic effect. As shown before, the fall of the signal above the gap may be due to free carrier contribution. Their participation should also be considered when generating the signal in the defective subsurface layer. This assumption can be supported by the most recent results obtained by Aleksić et al. [31]. In their article, the dependence of the photoacoustic signal on the frequency of the excitation beam was used to analyze the influence of thin transparent foil on the thermal and elastic properties of a two-layer sample consisting of a silicon substrate and a thin TiO_2 foil. A thin layer can significantly affect the substrate's thermal state and increase the sample's bending. This is related to a slight change in the number of photogenerated carriers in the transparent layer, which strongly influences the temperature differences between the illuminated and non-illuminated sides of the sample [31]. The influence of the carriers in the defective layer probably extinguished the signal for the first sample etching procedure.

5. Conclusions

Using piezoelectric photothermal spectroscopy, a new $Cd_{1-x}Be_xTe$ material was investigated, and its basic optical parameters were determined (E_g primary and thermal diffusivity). The influence of the sample preparation method on the amplitude and phase of the photothermal signal was observed and interpreted. High-quality sample surfaces have yet to be achieved, and further work on surface treatment procedures is needed. Further research and development of an appropriate etching method are required to obtain a defect-free surface of the samples. The next step will be to measure the different etching mixtures and etching times. The extinction of the signal-etching procedure may have been caused by excessive etching time. Despite not achieving good surface quality, piezoelectric spectroscopy is a sensitive method and can help choose the appropriate surface treatment method to obtain the desired quality.

In this paper, we presented a preliminary interpretation of our experimental data using the modified Blonskij model. As not all courses of the amplitude and phase could be interpreted with the proposed theory, it is necessary to expand it. We have shown that considering only the thermoelastic effect for analyzing and simulating experimental spectra cannot fully reproduce them. Effects related to carriers should also be considered in the interpretation, both in terms of nonradiative bulk recombination of carriers that diffuse in the crystal and the nonradiative surface recombination of carriers.

Author Contributions: Methodology, D.M.K.; Investigation, J.Z., K.S., M.B., A.M., A.A. and D.M.K.; Writing—original draft, J.Z. All authors have read and agreed to the published version of the manuscript.

Funding: This research was funded by Emerging Fields Team IDUB of Nicolaus Copernicus University—Material Science and Technology, 2023–2026.

Institutional Review Board Statement: Not applicable.

Data Availability Statement: The data presented in this study are available on request from the corresponding author.

Conflicts of Interest: The authors declare no conflict of interest.

References

1. Schlesinger, T.; Toney, J.; Yoon, H.; Lee, E.; Brunett, B.A.; Franks, L.; James, R.B. Cadmium zinc telluride and its use as a nuclear radiation detector material. *Mater. Sci. Eng.* **2001**, *32*, 103–189. [CrossRef]
2. Choi, H.; Jeong, M.; Kim, H.S.; Kim, Y.S.; Ha, J.H.; Chai, J.-S. Growth and Fabrication Method of CdTe and Its Performance As a Radiation Detector. *J. Korean Phys. Soc.* **2015**, *66*, 27–30. [CrossRef]
3. Szeles, C.; Cameron, S.E.; Ndap, J.O.; Chalmers, W. Advances in the High-pressure Crystal Growth Technology of Semi-insulating CdZnTe for Radiation Detector Applications. *IEEE Trans. Nucl. Sci.* **2002**, *49*, 2424–2428. [CrossRef]
4. Del Sordo, S.; Abbene, L.; Caroli, E.; Mancini, A.M.; Zappettini, A.; Ubertini, P. Progress in the Development of CdTe and CdZnTe Semiconductor Radiation Detectors for Astrophysical and Medical Applications. *Sensors* **2009**, *9*, 3491–3526. [CrossRef]
5. Partovi, A.; Millerd, J.; Garmire, E.M.; Ziari, M.; Steier, W.H.; Trivedi, S.B.; Klein, M.B. Photorefractivity at 1.5 μm in CdTe:V. *Appl. Phys. Lett.* **1990**, *57*, 846–848. [CrossRef]
6. Sen, S.; Liang, C.S.; Rhiger, D.R.; Stannard, J.E.; Arlinghaus, H.F. Reduction of CdZnTe substrate defects and relation to epitaxial HgCdTe quality. *J. Electron. Mater.* **1996**, *25*, 1188–1195. [CrossRef]
7. Szeles, C.; Soldner, S.A.; Vydrin, S.; Graves, J.; Bale, D.S. CdZnTe Semiconductor Detectors for Spectroscopic X-ray Imaging. *IEEE Trans. Nucl. Sci.* **2008**, *55*, 572–582. [CrossRef]
8. Zhoua, C.; Yang, J.; Yu, H.; Xu, C.; Shi, Y. Study of the extended defects in CdZnTe crystal. *J. Cryst. Growth* **2020**, *544*, 125725.
9. Gu, J.; Shen, Y.; Wen, D.; Huang, J.; Lai, J.; Gu, F.; Cao, M.; Wang, L.; Min, J. Marked effects of Al-rich AlN transition layers on the performance of CdZnTe films for solar-blind photodetector. *Vacuum* **2021**, *193*, 110539. [CrossRef]
10. Li, Y.; Zha, G.; Guo, Y.; Xi, S.; Xu, L.; Yu, H.; Jie, W. Effects of deep-level traps on the transport properties of high-flux X-ray CdZnTe detectors. *Mater. Sci. Semicond. Process.* **2021**, *133*, 105974.
11. Aqariden, F.; Tari, S.; Nissanka, K.; Li, J.; Kioussis, N.; Pimpinella, R.E.; Dobrowolska, M. Influence of surface polishing on the structural and electronic properties of CdZnTe surfaces. *J. Electron. Mater.* **2012**, *41*, 2893–2898. [CrossRef]
12. Zheng, Q.; Dierre, F.; Crocco, J.; Carcelen, V.; Dieguez, E. Influence of surface preparation on CdZnTe nuclear radiation detectors. *Appl. Surf. Sci.* **2011**, *257*, 742–8746. [CrossRef]
13. Tari, S.; Aqariden, F.; Chang, Y.; Grein, C.; Li, J.; Kioussis, N. Impact of surface treatment on the structural and electronic properties of polished CdZnTe surfaces for radiation detectors. *J. Electron. Mater.* **2013**, *42*, 3252–3258. [CrossRef]
14. Hossain, A.; Bolotnikov, A.E.; Camarda, G.S.; Cui, Y.; Jones, D.; Hall, J.; Kim, K.H.; Mwathi, J.; Tong, X.; Yang, G.; et al. Novel approach to surface processing for improving the efficiency of CdZnTe detectors. *J. Electron. Mater.* **2014**, *43*, 2771–2777. [CrossRef]
15. Song, B.; Zhang, J.; Liang, X.; Zhao, S.; Min, J.; Shi, H.; Lai, J.; Wang, L. Effects of the inductively coupled Ar plasma etching on the performance of (111) face CdZnTe detector. *Mater. Sci. Semicond. Process.* **2020**, *109*, 104929. [CrossRef]
16. Zhang, Z.; Wang, B.; Zhou, P.; Kang, R.; Zhang, B.; Guo, D. A novel approach of chemical mechanical polishing for cadmium zinc telluride wafers. *Sci. Rep.* **2016**, *6*, 26891. [CrossRef]
17. Min, J.; Li, H.; Wang, L.; Zhang, J.; Liang, X.; Liu, W.; Sun, X.; Wang, D.; Yuan, Z.; Wei, G.; et al. The effect of hydrogen plasma treatment on the Au–CdZnTe interface. *Surf. Coat. Technol.* **2013**, *228*, 97–100. [CrossRef]
18. Zakrzewski, J.; Maliński b, M.; Bachiri, A.; Strzałkowski, K. Photothermal determination of the optical and thermal parameters of $Cd_xZn_{1-x}Se$ mixed crystals. *Mater. Sci. Eng. B* **2021**, *271*, 115305. [CrossRef]
19. Maliński, M.; Zakrzewski, J.; Strzałkowski, K. Numerical Analysis of Piezoelectric Spectra of $Zn_{1-x-y}Be_xMn_ySe$ Mixed Crystals. *Int. J. Thermophys.* **2007**, *28*, 229–316. [CrossRef]

20. Zakrzewski, J.; Maliński, M.; Strzałkowski, K. Influence of the Surface Mechanical Treatment on the Photothermal Piezoelectric Spectra of ZnSe Crystals. *Int. J. Thermophys.* **2012**, *33*, 1228–1238. [CrossRef]
21. Singh, D.; Strzałkowski, K.; Abouais, A.; Alaoui-Belghiti, A. Study of the Thermal Properties and Lattice Disorder Effects in CdTe–Based Crystals: CdBeTe, CdMnTe, and CdZnTe. *Crystals* **2022**, *12*, 1555. [CrossRef]
22. Firszt, F.; Łegowski, S.; Meczynska, H.; Szatkowski, J.; Paszkowicz, W.; Godwod, K. Growth, and characterization of $Zn_{1-x}Be_xSe$ mixed crystals. *J. Cryst. Growth* **1998**, *184–185*, 1335. [CrossRef]
23. Gorgol, M.; Zaleski, R.; Kierys, A.; Kamiński, D.; Strzałkowski, K.; Fedus, K. Positron lifetime spectroscopy of defect structures in $Cd_{1-x}Zn_xTe$ mixed crystals grown by vertical Bridgman–Stockbarger method. *Acta Cryst.* **2021**, *B77*, 515–525.
24. Downs, R.T.; Hall-Wallace, M. The American Mineralogist Crystal Structure Database. *Am. Miner.* **2003**, *88*, 247–250.
25. Zakrzewski, J.; Maliński, M.; Strzałkowski, K.; Firszt, F.; Łegowski, S.; Męczyńska, H. Influence of Surface Preparation for Different Groups of A2B6 Mixed Crystals. *Int. J. Thermophys.* **2010**, *31*, 208–217. [CrossRef]
26. Maliński, M.; Zakrzewski, J.; Strzałkowski, K.; Łegowski, S.; Firszt, F.; Męczyńska, H. Piezoelectric photoacoustic spectroscopy of surface states of $Zn_{1-x-y}Be_xMg_ySe$ mixed crystals. *Surf. Sci.* **2009**, *603*, 131–137. [CrossRef]
27. Kuwahata, H.; Muto, N.; Uehara, F.; Matsumori, T. Comparison of Annealing Behavior in Photoacoustic Signal Intensity of Si+ Implanted InP by Microphone an Piezoelectric Transducer Method. *AIP Conf. Proc.* **1999**, *463*, 250–252.
28. Matsumori, T.; Uchida, M.; Yoshinaga, H.; Kawai, J.; Izumi, T.; Uehara, F. Photoacoustic Characterization Of Ion-Implanation Damage in Silicon. In *Phoacoustic and Photothermal Phenomena III*; Springer Series in Optical Sciences; Springer: Berlin, Germany, 1992; Volume 69, pp. 357–359.
29. Blonskij, I.; Tkhoryk, V.; Shendeleva, M. Thermal diffusivity of solids determination by photoacoustic piezoelectric technique. *J. Appl. Phys.* **1996**, *79*, 3512–3516. [CrossRef]
30. Zakrzewski, J.; Maliński, M.; Chrobak, Ł.; Pawlak, M. Comparison of Theoretical Basics of Microphone and Piezoelectric Photothermal Spectroscopy of Semiconductors. *Int. J. Thermophys.* **2017**, *38*, 2. [CrossRef]
31. Aleksić, S.M.; Markushev, D.K.; Markushev, D.D.; Pantić, D.S.; Lukić, D.V.; Popović, M.N.; Galović, S.P. Photoacoustic Analysis of Illuminated Si-TiO$_2$ Sample Bending Along the Heat-Flow Axes. *Silicon* **2022**, *14*, 9853–9861. [CrossRef]

Article

The Adhesive Force Measurement between Single μLED and Substrate Based on Atomic Force Microscope

Jie Bai [1], Pingjuan Niu [2,*], Shinan Cao [1] and Qiang Liu [3]

[1] School of Mechanical Engineering, Tiangong University, Tianjin 300387, China
[2] School of Electronic and Information Engineering, Tiangong University, Tianjin 300387, China
[3] Institute of Precision Electromagnetic Equipment and Advanced Measurement Technology, Beijing Institute of Petrochemical Technology, Beijing 102617, China
* Correspondence: niupingjuan@tiangong.edu.cn

Abstract: Compared with traditional liquid crystal and organic light emitting diode (OLED), micro light emitting diode (μLED) has advantages in brightness, power consumption, and response speed. It has important applications in microelectronics, micro-electro-mechanical systems, biomedicine, and sensor systems. μLED massive transfer method plays an important role in these applications. However, the existing μLED massive transfer method is faced with the problem of low yield. To better transfer the μLED, the force value detached from the substrate needs to be measured. Atomic force microscope (AFM) was used to measure the force of a single μLED when it detached from the substrate. The μLED was glued to the front of the cantilever. When a single μLED was in contact with or detached from the Polydimethylsiloxane (PDMS), the maximum pull-off force can be obtained. The force at different peel speeds and preload was measured, and the experimental results show that the separation force between a single μLED and PDMS substrate is not only related to the peel speeds, but also related to the preload. The force values under different peel speeds and preload were measured to lay a theoretical foundation for better design of μLED massive transfer system.

Keywords: AFM; μLED; mass transfer; adhesive force measurement

Citation: Bai, J.; Niu, P.; Cao, S.; Liu, Q. The Adhesive Force Measurement between Single μLED and Substrate Based on Atomic Force Microscope. *Appl. Sci.* **2022**, *12*, 9480. https://doi.org/10.3390/app12199480

Academic Editor: Andrey Miroshnichenko

Received: 30 July 2022
Accepted: 19 September 2022
Published: 21 September 2022

Publisher's Note: MDPI stays neutral with regard to jurisdictional claims in published maps and institutional affiliations.

1. Introduction

The main advantage of micro light emitting diode (μLED) is that each LED can be controlled and driven independently, leading to excellent power consumption, brightness, resolution, contrast, heat dissipation, and other characteristics [1–3], and μLED has important applications in microelectronics [4], biomedicine [5], and sensor systems [6].

μLED needs to be transferred to the circuit substrate in practical application. Currently, the most popular transfer method is the stamp method, which adjusts the adhesion force by adjusting the peeling parameters of Polydimethylsiloxane (PDMS) to complete the μLED pickup and release. The design of stamp is one of the key technologies of μLED transfer printing [7–9]. A comprehensive understanding of the adhesion between μLED and PDMS is needed to achieve more efficient and high-yield transfer. Therefore, it is very important to measure the adhesion between the μLED and the substrate under different peel speeds and preload.

Recently, many scholars have studied the adhesion between μLED and substrate. Tian Yu et al. found that the peel angle can regulate the adhesion and friction through a theoretical model, which is the mechanism of gecko's strong adhesion and fast separation [10]. Xu Quan, Rogers et al. found that peel speed is an important factor affecting adhesion [11]. Rogers optimized the ground geometry of PDMS by using a sharp substrate, and the strength of adhesion can be switched from strong to weak in a reversible manner by more than three orders of magnitude [9].

Chang Dong Yeo developed an instrument based on capacitive force sensor to measure the dynamic adhesion between rough surfaces [12]. Min sock Kim proposed a new instrument to measure the dynamic adhesion of the interaction surface on the flexible substrate, and proposed an optimal adhesion control strategy based on the analysis of adhesion [13]. In 2017, Lindsay Vasilak used strain gauge load cell to measure the normal adhesive force of OLED [14]. Chang-Dong Yeo used a high-resolution, high-dynamic bandwidth capacitive force transducer and two piezoelectric actuators to measure adhesive pull-off forces between nominally flat rough silicon surfaces under various dynamic conditions [15]. Jaeho Lee used Atomic Force Microscope (AFM) to measure the adhesion between the colloidal probe and silicon wafer. Two spherical colloids made of silicon dioxide and gold that were attached to an AFM cantilever were approached to and retracted from a silicon wafer specimen [16].

However, at present, none has been found in the literature regarding adhesion test between a single μLED and the substrate and most of the experiments demonstrate the adhesion between a large area of PDMS and μLED array, since it is hard to attach a single μLED to the force sensor. In this paper, based on AFM, the adhesion between a single μLED and the substrate was measured using cantilever, and the relationship between peel speeds, preloads, and adhesion was evaluated [17,18]. In Section 2, the theoretical relationship between pull-off forces to peel speeds and preload will be deduced. In Section 3, the theoretical results will be verified experimentally.

2. The Theoretical Relationship between Pull-Off Forces to Peel Speed and Preload

2.1. The Image of μLED

The size of μLED is generally smaller than 100 μm [19,20], as shown in Figure 1. A scanning electron microscope (SEM) image of μLED is shown in Figure 1A. The surface of the μLED array obtained by optical microscope is shown in Figure 1B.

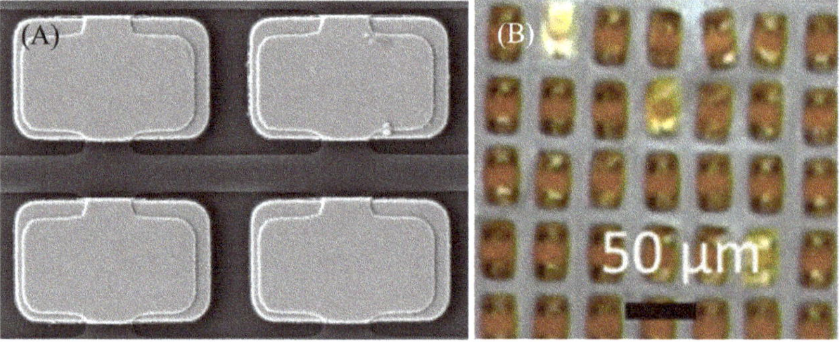

Figure 1. Image of μLED. (**A**) Scanning electron microscope (SEM) image of μLED. (**B**) The surface of the μLED array obtained by optical microscope.

2.2. The Theoretical Relationship between Pull-Off Forces and Peel Velocity

The relationship between peel speed and pull-off force has also been extensively studied [21], which can be expressed as:

$$G_c(v) = G_0[1 + (\frac{v}{v_0})^k]$$

(1)

where G_0 is the critical energy release rate and corresponding detaching speed v_0 approaches zero, v is the peel speed, and the exponent k is a parameter that can be determined from experiments. The power–law relationship (Equation (1)) has been found applicable to low or high peel speed obtained from metal/polymer and polymer/polymer interfaces at various temperatures.

2.3. The Theoretical Relationship between Pull-Off Forces and Preload

According to the Hertzian contact theory, the actual contact occurs only on a small part of the apparent area due to the surface roughness when two solid surfaces are in contact. The size and distribution of the zone of contact exert a decisive influence on friction and wear. The shape of the rough peaks on the actual contact surface is usually elliptical. Since the size of the contact area of the ellipsoid is much smaller than its radius of curvature, the rough peak can be approximately regarded as a sphere. The contact of two flat surfaces can be regarded as a series of uneven spheres. The contact between two elastomer can be converted into the contact between an elastic sphere with equivalent radius of curvature R and equivalent modulus of elasticity E and a rigid smooth surface.

When μLED contacts with PDMS, the Young's modulus of PDMS is much lower than that of μLED, so it can be considered as elastic contact. When the two rough peaks contact each other, the normal deformation δ is produced under the action of load W, which makes the shape of the elastic sphere change from dotted line to solid line. The actual contact area is a circle of radius a, as shown in Figure 2. The relationship between load and contact area is given by Equation (2) [22].

$$
\begin{cases}
\delta = \left(\dfrac{9W^2}{16E^{*2}R}\right)^{1/3} \\[2mm]
a = \left(\dfrac{3WR}{4E^*}\right)^{1/3} \\[2mm]
W = \dfrac{4}{3}E^* R^{1/2}\delta^{3/2}
\end{cases}
\tag{2}
$$

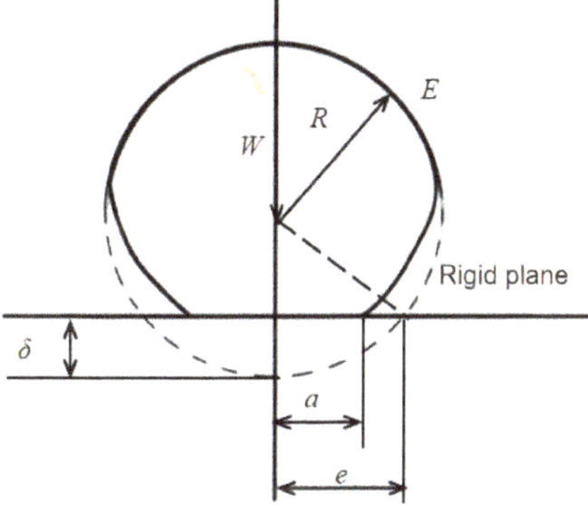

Figure 2. Diagram of single peak elastic contact.

The ideal rough surface is composed of many orderly rough peaks with the same curvature radius and height, and the load and deformation of each peak are exactly the same and independent from each other. However, the rough peak height of the actual contact surface is randomly distributed in general, so the contact peak should be calculated according to the probability. The contact condition of two rough surfaces is shown in Figure 3.

Their contact can be converted into the situation where one smooth rigid surface touches another rough elastic surface. Since the surface of μLED is very smooth, while the surface of PDMS is quite the opposite, this assumption is consistent with reality.

When the distance between the center lines is h, only the part of the contour height $z > h$ contacts with. In the probability density distribution curve, the shading area of the $z > h$ part is the surface contact probability, that is [23]

$$
P(z > h) = \int_h^{\infty} \psi(z)\,\mathrm{d}z
\tag{3}
$$

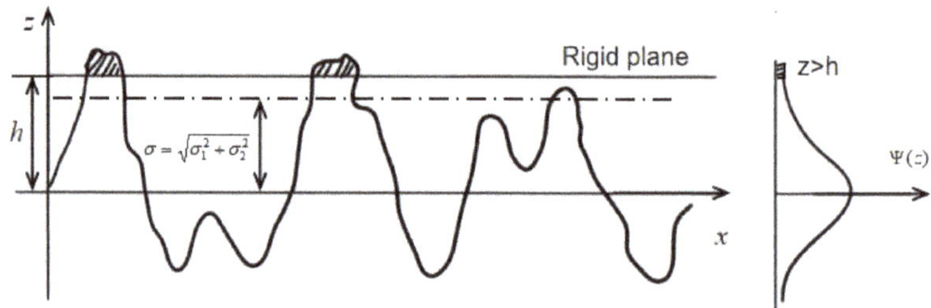

Figure 3. Contact of rough surfaces. The root mean square values of the roughness of the two surfaces are respectively σ_1 and σ_2, h is the distance between the center lines, z is the part of the contour height $z > h$ contacts with, and $\Psi(z)$ is the probability of the surface contact.

If the number of peaks on the rough surface is n, the number of peaks participating in the contact, m, is given by [23]:

$$m = n \int_h^\infty \psi(z)\mathrm{d}z \tag{4}$$

The normal phase deformation of each contact peak is z-h. From Equation (2), the actual contact area A is given by [23]:

$$A = m\pi R(z - h) = n\pi R \int_h^\infty (z - h)\psi(z)\mathrm{d}z \tag{5}$$

The total load W is supported by the contact peak as [23]:

$$W = \frac{4}{3}mE^*R^{1/2}(z - h)^{3/2} = \frac{4}{3}nE^*R^{1/2}\int_h^\infty (z - h)^{3/2}\psi(z)\mathrm{d}z \tag{6}$$

Usually, the contour height of the actual surface follows a Gaussian distribution [24], in which most of region near the z-score approximates an exponential distribution. Suppose that $\psi(z) = \exp(-z/\sigma)$, we get:

$$m = n\sigma \exp(-h/\sigma) \tag{7}$$

$$A = \pi n R\sigma^2 \exp(-h/\sigma) \tag{8}$$

$$W = \frac{3}{4}nE^*R^{1/2}\sigma^{3/2} \exp(-h/\sigma) \tag{9}$$

From the above equations, it can be derived that W is proportional to A and W is proportional to m. Thus, the actual contact area and the number of contact peaks have a linear relationship with the load in the elastic contact state of the two rough surfaces. Separating μLEDs from PDMSs creates two new interfaces, and the force value F_{cr} required for this process is obtained as [10]:

$$F_{cr} = A\gamma \tag{10}$$

where γ is the viscosity coefficient of the two surfaces. From Equations (8)–(10), it can be concluded that the adhesive force increases with the increase of preload.

3. The Experimental Results and Discussion

3.1. Experimental Steps

To measure the adhesion between a single μLED and the substrate, a cantilever measurement scheme is adopted, and the specific steps are shown in Figure 4A–C.

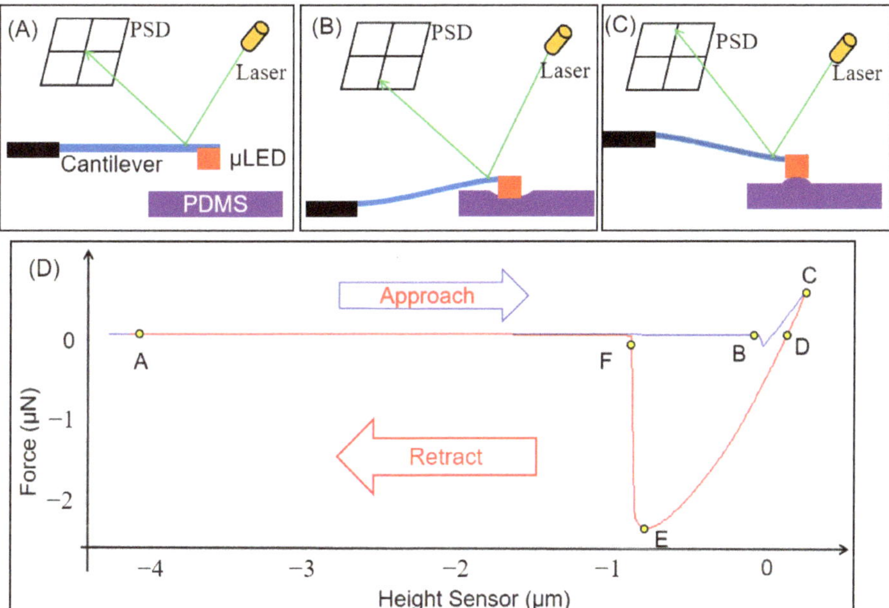

Figure 4. The adhesion measurement results based on cantilever between µLED and substrate. (**A–C**) Measurement steps. (**A**) The initial state. (**B**) The loading status. (**C**) The reverse motion. (**D**) Typical single measurement results.

Step 1: Apply glue to the tip of the tipless cantilever with a stiffness of 5.1 N/m.

Step 2: Move the cantilever above a µLED.

Step 3: Lower the cantilever to contact the µLED and wait for the glue (UV photoresist) to solidify.

Step 4: Raise the cantilever to make the µLED separate from the base.

Step 5: Move µLED above the PDMS substrate.

Step 6: Measure the relevant force value at different peel speeds and preload.

A typical adhesion–depth curve on a single µLED with a flexible PDMS substrate (1:10 mixing ratio) measured by AFM is shown in Figure 4D. The tip of the cantilever is controlled at a speed of 10 µm/s. The x-axis is the displacement of the µLED. The measurement process is divided into two segments according to the direction of cantilever movement: approach (red line in Figure 4D) and retract (blue line in Figure 4D). The y-axis is force between the µLED and PDMS.

The AFM has been well calibrated using thermal method. The relationship between the force acted on cantilever and PSD output has been obtained before measuring the adhesion force.

The µLED on the cantilever was moved above a substrate PDMS, as shown in Figure 4A. Figure 4B is in a loading status. The µLED is pressed on the PDMS and continuously moved through the precision stage. The laser spot moves as the cantilever bends. The cantilever will not stop until the pressure equals the set preload, as shown in section BC in Figure 4D. Figure 4C is in reverse motion. With the reverse movement, the pressure of µLED on the PDMS substrate becomes smaller and smaller until the pressure reaches 0, as shown in the CD section.

Dynamic jumping behavior during approach (such as the BC segment) and measured jumping behavior during return (CF) were measured.

As can be seen from Figure 4D, the measurement process can be divided into four stages according to the contact status: (I) pressure down to contact with PDMS, (II) pressure to the maximum to reach preload, (III) reverse movement, and (IV) separation from PDMS.

I—Initial state: the cantilever is moved by a precision stage and is not in contact with the PDMS, as shown in Section AB.

II—Loading status: μLED contact PDMS. The μLED is pressed on the PDMS and continuously moved through the precision stage. The laser spot moves as the cantilever bends. The cantilever will not stop until the pressure equals the set preload, as shown in section BC.

III—Reverse motion: With the reverse movement, the pressure of μLED on the PDMS substrate becomes smaller and smaller until the pressure reaches 0, as shown in the CD section. As the reverse motion continues, the PDMS deforms due to the tension between the μLED and PDMS. At this point, the elastic force of the cantilever acting on μLED is less than the critical adhesion force of PDMS, as shown in section DE.

IV—Exit stage: The elastic force of the cantilever on μLED is greater than the critical adhesion force. A sudden jump in the position sensitive device (PSD) voltage output can be observed, as shown in the EF section.

The maximum pull-off force can be defined as the minimum force of the force-depth curve, as shown in Figure 4D.

3.2. Measurement of Adhesion under Different Detaching Velocities and Preload

As shown in Figure 5, the maximum pull-off force was measured at different peel velocities (detaching velocity) varying from 10 μm/s to 300 μm/s, in which high peel speed (300 μm/s) resulted in strong adhesion, while low peel speed (10 μm/s) resulted in weak adhesion. Obviously, there is a strong correlation (proportional relationship) between the maximum pull-off force and the peel velocity.

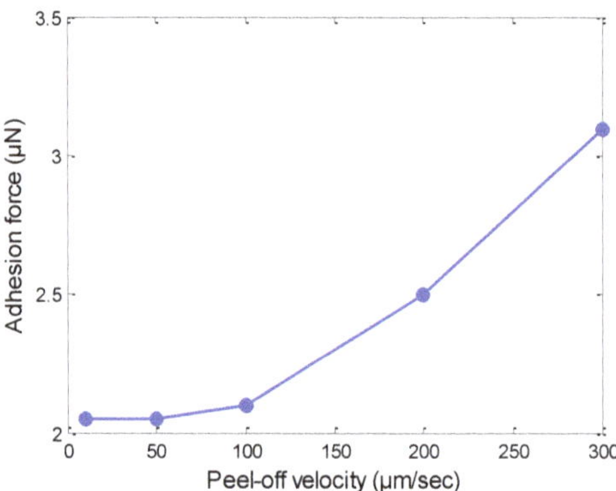

Figure 5. The measured maximum pull-off forces with respect to peel velocity.

We measured the maximum adhesion force at different preload from 0.5 to 3 μN (the peel speed was fixed at 10 μm/s). The proportional relationship between the maximum pull-off force and the preload is shown in Figure 6.

The experimental results show that the preload has a great influence on the adhesion, which is different from the previous research: "Unlike the effects of material property of PDMS, the maximum pull-off force has similar value regardless of the initial indentation force between the tip and the flexible substrate". Our theoretical result is consistent with our experimental result but different from the literature.

It Is hard to compare the experimental results to results from Equations (8)–(10). The equations show a positive proportion relationship between contact area and the maximum pull-off force. However, the real contact area between the μLED and PDMS or other substrates could not be measured. Therefore, it is impossible to directly compare the quantity of theoretical value and experimental value absolutely, but only relatively.

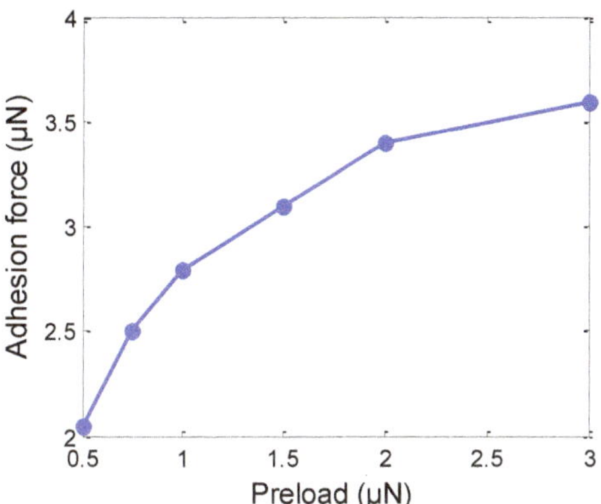

Figure 6. Results for the maximum pull-off force at different preload.

4. Conclusions

In this paper, the adhesion force between the μLED and substrate at different peel speeds and preload was measured by AFM. The experimental results show that the separation force between a single μLED and PDMS substrate is not only related to the peel speed, but also related to the preload. Although it is hard to directly compare the absolute quantity of theoretical value and experimental value, the results find a new way to design an apparatus for μLED transfer printing. Future research is required to reversibly change adhesion strength between strong and weak modes by more than two orders of magnitude so that the system can be applied in transfer printing. We will focus on the design of a novel substrate to achieve this target. This system would have broader impacts in transfer printing.

Author Contributions: J.B. and P.N. designed the manuscript, wrote and analyzed the data; J.B. designed the figures for the manuscript and performed the experiments, data collection and/or statistical analysis; P.N., S.C. and Q.L. revised the manuscript. All authors have read and agreed to the published version of the manuscript.

Funding: This research received financial support by the Scientific Research Program of Tianjin Education Commission (No.2019ZD08).

Institutional Review Board Statement: Not applicable.

Informed Consent Statement: Not applicable.

Data Availability Statement: The data that support the findings of this study are available from the corresponding author, P.J.N., upon reasonable request.

Conflicts of Interest: The authors declare that the research was conducted in the absence of any commercial or financial relationships that could be construed as a potential conflict of interest.

References

1. Virey, E.H.; Baron, N. Status and prospects of microLED displays. *SID Symp. Dig. Tech. Pap.* **2018**, *49*, 593–596. [CrossRef]
2. Lianga, C.; Wanga, F.; Huoa, Z.; Shia, B.; Tiana, Y.; Zhaoa, X.; Zhang, D. Pull-off force modeling and experimental study of PDMS stamp considering preload in micro transfer printing. *Int. J. Solids Struct.* **2020**, *193–194*, 134–140. [CrossRef]
3. Zhu, G.; Liu, Y.; Ming, R.; Shi, F.; Cheng, M. Mass transfer, detection and repair technologies in micro-LED displays. *Sci. China Mater.* **2022**, *65*, 2128–2153. [CrossRef]
4. Kamarei, Z.B. Analysis for science librarians of the 2014 Nobel Prize in physics: Invention of efficient blue-light-emitting diodes. *Sci. Technol. Libr.* **2015**, *34*, 19–31. [CrossRef]

5.   Park, S.-I.; Xiong, Y.; Kim, R.-H.; Elvikis, P.; Meitl, M.; Kim, D.-H.; Wu, J.; Yoon, J.; Yu, C.-J.; Liu, Z.; et al. Printed assemblies of inorganic light-emitting diodes for deformable and semitransparent displays. *Science* **2009**, *325*, 977–981. [CrossRef]

6.   Cok, R.S.; Meitl, M.; Rotzoll, R.; Melnik, G.; Fecioru, A.; Trindade, A.J.; Raymond, B.; Bonafede, S.; Gomez, D.; Moore, T.; et al. Inorganic light-emitting diode displays using micro-transfer printing. *J. Soc. Inf. Disp.* **2017**, *25*, 589–609. [CrossRef]

7.   Kim, T.-I.; Jung, Y.H.; Song, J.; Kim, D.; Li, Y.; Kim, H.-S.; Song, I.-S.; Wierer, J.J.; Pao, H.A.; Huang, Y.; et al. High-efficiency, microscale GaN light-emitting diodes and their thermal properties on unusual substrates. *Small* **2012**, *8*, 1643–1649. [CrossRef]

8.   Carlson, A.; Bowen, A.M.; Huang, Y.; Nuzzo, R.G.; Rogers, J.A. Transfer printing techniques for materials assembly and micro/nanodevice fabrication. *Adv. Mater.* **2012**, *24*, 5284–5318. [CrossRef]

9.   Kim, S.; Wu, J.; Carlson, A.; Jin, S.H.; Kovalsky, A.; Glass, P.; Liu, Z.; Ahmed, N.; Elgan, S.L.; Chen, W.; et al. Microstructured elastomeric surfaces with reversible adhesion and examples of their use in deterministic assembly by transfer printing. *Proc. Natl. Acad. Sci. USA* **2010**, *107*, 17095–17100. [CrossRef]

10. Tian, Y.; Pesika, N.; Zeng, H.; Rosenberg, K.; Zhao, B.; McGuiggan, P.; Autumn, K.; Israelachvili, J. Adhesion and friction in gecko toe attachment and detachment. *Proc. Natl. Acad. Sci. USA* **2006**, *103*, 19320–19325. [CrossRef]

11. Xu, Q.; Wan, Y.; Hu, T.S.; Liu, T.X.; Tao, D.; Niewiarowski, P.H.; Tian, Y.; Liu, Y.; Dai, L.; Yang, Y.; et al. Robust Self-cleaning and Micromanipulation Capabilities of Nano-pads of Gecko Spatulae and their Bio-mimics. *Nat. Commun.* **2015**, *6*, 8949. [CrossRef]

12. Yeo, C.; Lee, S.; Polycarpou, A.A. Dynamic adhesive force measurements under vertical and horizontal motions of interacting rough surfaces. *Rev. Sci. Instrum.* **2008**, *79*, 015111. [CrossRef]

13. Kim, M.S.; Park, J.; Choi, B. Measurement and analysis of micro-scale adhesion for efficient transfer printing. *J. Appl. Phys.* **2011**, *110*, 024911. [CrossRef]

14. Vasilak, L.; Halim, S.M.T.; Gupta, H.D.; Yang, J.; Kamperman, M.; Turak, A. Statistical Paradigm for Organic Optoelectronic Devices: Normal Force Testing for Adhesion of Organic Photovoltaics and Organic Light-Emitting Diodes. *ACS Appl. Mater. Interfaces* **2017**, *9*, 13347–13356. [CrossRef]

15. Yeo, C.; Lee, J.; Polycarpou, A.A. Dynamic adhesive forces in rough contacting bodies including normal and sliding conditions. *J. Adhes. Sci. Technol.* **2012**, *26*, 2709–2718. [CrossRef]

16. Lee, J.; Maharjan, J.; He, M.; Yeo, C. Effects of system dynamics and applied force on adhesion measurement in colloidal probe technique. *Int. J. Adhes. Adhes.* **2015**, *60*, 109–116. [CrossRef]

17. Zheng, Y.; Song, L.; Hu, G.; Zhao, M.; Tian, Y.; Zhang, Z.; Fang, F. Improving environmental noise suppression for micronewton force sensing based on electrostatic by injecting air damping. *Rev. Sci. Instrum.* **2014**, *85*, 055002. [CrossRef]

18. Lee, M.H.; Lim, N.; Ruebusch, D.J.; Jamshidi, A.; Kapadia, R.; Lee, R.; Seok, T.J.; Takei, K.; Cho, K.Y.; Fan, Z.; et al. Roll-to-roll anodization and etching of aluminum foils for high-throughput surface nanotexturing. *Nano Lett.* **2011**, *11*, 3425–3430. [CrossRef]

19. Liang, C.; Wang, F.; Huo, Z.; Shi, B.; Tiana, Y.; Zhang, D. Adhesion performance study of a novel microstructured stamp for micro-transfer printing. *Soft Matter.* **2021**, *17*, 4989. [CrossRef]

20. Qiu, P.; Zhu, S.; Jin, Z.; Zhou, X.; Cui, X.; Tian, P. Beyond 25 Gbps optical wireless communication using wavelength division multiplexed LEDs and micro LEDs. *Opt. Lett.* **2022**, *47*, 317–320. [CrossRef]

21. Feng, X.; Meitl, M.A.; Bowen, A.M.; Huang, Y.; Nuzzo, R.G.; Rogers, J.A. Competing fracture in kinetically controlled transfer printing. *Langmuir* **2007**, *23*, 12555–12560. [CrossRef] [PubMed]

22. Shizhu, W. *Principles of Tribology*, 2nd ed.; Tsinghua University Press: Beijing, China, 2017.

23. Bowden, F.P.; Tabor, D. *The Friction and Lubrication of Solid*; Clarenden Press: Oxford, UK, 1964.

24. Johnson, K.L. *Contact Mechanics*; Cambridge University Press: London, UK, 1985.

applied sciences

Article

Low-Frequency Terahertz Photonic Crystal Waveguide with a Lilac-Shaped Defect Based on Stereolithography 3D Printing

Jia Shi [1,*], Yiyun Ding [1], Longhuang Tang [2], Xiuyan Li [1], Hua Bai [1], Xianguo Li [1], Wei Fan [1], Pingjuan Niu [1], Weiling Fu [3], Xiang Yang [3,*] and Jianquan Yao [4]

[1] Tianjin Key Laboratory of Optoelectronic Detection Technology and System, School of Electronic and Information Engineering, Tiangong University, Tianjin 300387, China
[2] Institute of Fluid Physics, China Academy of Engineering Physics, Mianyang 621900, China
[3] Department of Laboratory Medicine, Southwest Hospital, Third Military Medical University (Army Medical University), Chongqing 400038, China
[4] Key Laboratory of Opto-Electronics Information Technology (Ministry of Education), School of Precision Instruments and Opto-Electronic Engineering, Tianjin University, Tianjin 300072, China
* Correspondence: shijia@tiangong.edu.cn (J.S.); yangxiang@tmmu.edu.cn (X.Y.)

Abstract: Terahertz (THz) photonic crystal (PC) waveguides show promise as an efficient and versatile waveguiding platform for communication, sensing, and imaging. However, low-frequency THz PC waveguides with a low-cost and easy fabrication remain challenging. To address this issue, a THz PC waveguide with a lilac-shaped defect has been designed and fabricated by 3D printing based on stereolithography (SLA). The reflection and transmission characteristics of the proposed waveguide have been analyzed using the finite difference frequency domain (FDFD) method. The waveguide spectral response is further optimized by changing the distance of the lilac-shaped resonant cavities. Consistent with the results of numerical modeling, the measured results show that the waveguide performs a resonant reflection in the region of 0.2 to 0.3 THz and low-pass transmission in the 6G mobile communication window. Furthermore, in order to characterize the performance of the proposed waveguide, parameters have been analyzed, including the Q factor, resonant frequency, and bandwidth. This work supplies a novel pathway for the design and fabrication of a low-frequency THz PC waveguide with potential applications in communication, sensing, and imaging.

Keywords: terahertz; photonic crystal waveguide; 3D printing

Citation: Shi, J.; Ding, Y.; Tang, L.; Li, X.; Bai, H.; Li, X.; Fan, W.; Niu, P.; Fu, W.; Yang, X.; et al. Low-Frequency Terahertz Photonic Crystal Waveguide with a Lilac-Shaped Defect Based on Stereolithography 3D Printing. *Appl. Sci.* **2022**, *12*, 8333. https://doi.org/10.3390/app12168333

Academic Editor: Mira Naftaly

Received: 26 July 2022
Accepted: 16 August 2022
Published: 20 August 2022

Publisher's Note: MDPI stays neutral with regard to jurisdictional claims in published maps and institutional affiliations.

1. Introduction

Terahertz (THz) is electromagnetic radiation with a frequency in the range 0.1 to 10 THz, which lies in the gap between the microwave and infrared regions [1]. THz technology has been playing an increasingly important role in various fields, including wireless communication, security, biomedical applications, imaging, sensing, and spectroscopy [2–4]. Although THz waves have proven to be beneficial for many applications, most THz systems are based on free-space optics that are complex, delicate, and require frequent alignment [1]. To solve these issues, different types of THz waveguides have been proposed. Many of THz waveguides have been demonstrated for applications such as communication, sensing, and imaging. Notably, the THz band from 0.1 to 0.3 THz has been of significant research interest in recent years, as it is considered to be the main transmission band for 6G telecommunication; low-frequency THz waveguides working in the 6G mobile communication window are gaining immense demand [5]. Currently, the very limited design library of conventional waveguide structures substantially constraints their functionalities to mostly mere waveguiding. Researchers are still struggling to design and manufacture high-performance waveguides with a low-cost and easy fabrication [3].

A variety of waveguides have been explored, including metallic and dielectric waveguides [4,6–12]. Most metallic waveguides are only suitable for millimeter lengths [13], because of the strong trade-off between mode confinement and metallic loss [11,12]. Because of the existing surface plasmon polaritons (SPPs) [12], metallic PCs enable confinement of THz waves in the sub-millimeter scale. However, metallic PC waveguides are still inevitably accompanied by Ohmic losses, which limit the quality factors of resonance and compromise the efficiency of metallic-based devices [8]. In the past few years, a growing number of reports have shown that this problem could be solved by employing dielectric waveguides, which mostly rely on high-index and low-loss particles (such as silicon, SiO_2, and TiO_2) [7,9,10]. Consequently, interest in dielectric waveguides has increased with the application of dielectric antennas, resonators, polarizers, etc. [4,8–10,14]. Meanwhile, a number of researches have been focused on dielectric waveguides because of the wide variety of available materials and the greater flexibility of the waveguide design. Many dielectric waveguide structures, including planar [15], rectangular [16], circular [1], strip [17], and photonic crystal waveguides [10,18,19] have been investigated. Among them, photonic crystals can realize strong light confinement due to the nature of photonic band gaps, and show great potential in high-performance dielectric waveguides [10].

Currently, THz waveguides are mostly fabricated by photolithography, which requires multiple steps, including spin-coating, prebaking, the preparation of masks, exposure, etc. [20–22], leading to a long preparation cycle and a high cost [16,23]. In addition, most of these devices have a substrate, which significantly lowers the transmittance and causes internal interference. Therefore, it is of great significance that THz PC waveguides are fabricated in a simple, low-cost, and efficient way [16]. Furthermore, most of the related works are conducted at a microwave and millimeter wave; it remains challenging to obtain low-frequency THz PC waveguides with a low-cost and easy fabrication. Direct writing technology has been applied to create complex THz waveguides, such as microfluidic three-dimensional photonic crystals [24]. High-accuracy 3D printing based on stereolithography (SLA) [13,23,25–27] as an emerging technology shows great potential in high-performance dielectric waveguides, but it has not been applied in THz PC waveguides.

In this work, a THz PC waveguide with a lilac-shaped defect is demonstrated. This proposed waveguide is fabricated with a high-density photosensitive resin by 3D printing based on SLA. The reflection and transmission properties of the waveguide in the 6G mobile communication window have been investigated. The waveguide spectral response is optimized by changing the distance of the lilac-shaped resonant cavities. The waveguide performance in the range of 0.1 THz to 0.5 THz has been analyzed, including the Q factors, resonant frequency (f_R), full width half max (FWHM), −60 dB bandwidth, and loss.

2. Design and Simulations

In this part, a low-frequency THz PC waveguide with a lilac-shaped defect is designed, as shown in Figure 1a,b. The symmetrical petal hollow core is induced in the waveguide structure to broaden the operation frequency bandwidth. For efficiently guiding the terahertz wave, the structural parameters are optimized to design a suitable structure. The lattice structure of the periodically arranged unit cells consists of air holes on the *o-xy* plane, which are described by the following [28,29]

$$x(M,m) = aM \cos\left(\frac{2m\pi}{6M}\right) \tag{1}$$

$$y(M,m) = aM \sin\left(\frac{2m\pi}{6M}\right) \tag{2}$$

where a is the lattice constant, M is the number of the air hole rings, and m ($1 \leq m \leq 6M$) is the number of the air holes in the M_{th} ring. The lilac-shaped resonant cavities are formed by four larger symmetrical air holes at the center, with the first, second, and third rings of the periodically arranged unit holes removed. Each resonant cavity consists of an intersection of one circular air hole with a diameter of D and one square air hole with a length of L. The gap distance between four petals of the lilac-shaped resonant cavities is defined as g. Initial values of the structure parameters are a = 1000 μm, d = 0.8 a = 800 μm, D = 1500 μm,

$L = 750$ μm, and $g = 500$ μm. Meanwhile, the depth of the air holes in the z direction is much larger than a ($h >> a$).

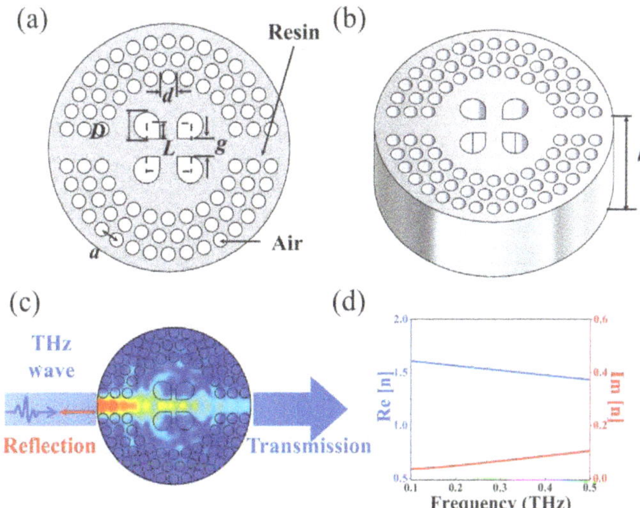

Figure 1. (**a**) The 2D structure of the THz PC waveguide. (**b**) The 3D structure of the THz PC waveguide. (**c**) The process of reflection and transmission characteristics of the THz PC waveguide. (**d**) The real and imaginary parts of the refractive index (RI) of the photosensitive resin.

In the simulation process, the optical characteristics of the THz PC waveguide were analyzed using COMSOL Multiphysics. The boundary condition was set to the scattering boundary condition to absorb energies. The physics-controlled mesh was applied to the model [29]. The process of the reflection and transmission characteristics of the proposed waveguide is shown in Figure 1c. The TE-polarized Gaussian form was injected into the waveguide from the left side and a monitor was set on the right side of the waveguide. Air holes were placed throughout the photosensitive resin background. In the simulation, the refractive index (RI) of the photosensitive resin used in the range of 0.1 to 0.5 THz were measured using the Terahertz Time-Domain Spectroscopy (THz-TDS) system (Menlo TeraSmart, Martinsried, Germany), which is shown in Figure 1d.

The gap distance (g) is an important parameter that can change the photonic bandgap of the photonic crystal; in particular, the guidance of the terahertz wave in the waveguide is caused by the photonic bandgap mechanism. Therefore, in order to explore the detailed functionality of the waveguide, the influence of the gap distance (g) has been simulated from 100 μm to 2000 μm in steps of 100 μm. The reflection map of the waveguide with different gap distances is shown in Figure 2a. As the gap distance increases from 100 μm to 2000 μm, the reflection loss performs a resonant dip in the frequency region of 0.2 to 0.3 THz. Correspondingly, as shown in Figure 2b, the transmission loss increases continually within 0.5 THz. For instance, when the gap distance is 1000 μm, the reflection loss (S11) and transmission coefficient (S21) are shown in Figure 2c. The band-stop behavior of the proposed waveguide is analyzed by the reflection loss. The resonant frequency of the band-stop is approximately at 0.23 THz with a transmission loss of about −44 dB. The transmission coefficient shows the low-pass behavior. To better understand the reflection and transmission effect, the simulated electric field patterns at different frequencies are analyzed in Figure 2d when g is 1000 μm. It is observed that the THz wave can smoothly pass through this waveguide within 0.2 THz and it is reflected mostly higher than 0.3 THz.

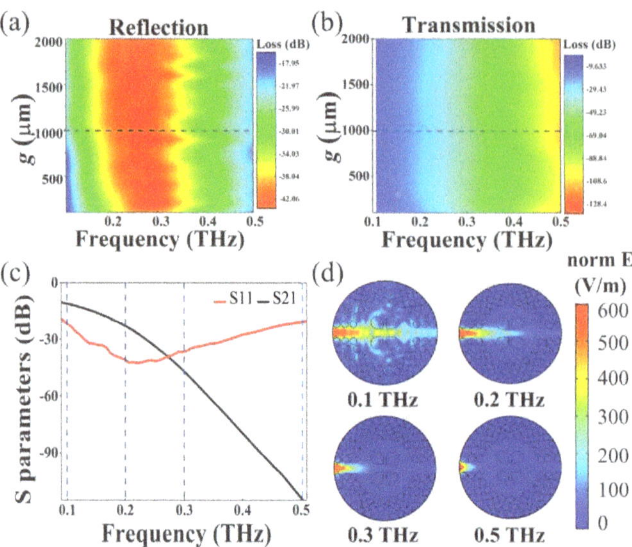

Figure 2. (**a**,**b**) The 2D contour maps of simulated reflection loss and transmission loss as a function of frequency with different gap distances (g), respectively. The black dashed line indicates g = 1000 µm (**c**) The simulated S parameters of the THz PC waveguide with g = 1000 µm. These blue dashed lines indicate the frequency at 0.1 THz, 0.2 THz, 0.3 THz, and 0.5 THz, respectively. (**d**) Electric field distributions of the THz PC waveguide at g = 1000 µm with different frequencies (0.1 THz, 0.2 THz, 0.3 THz, 0.5 THz, respectively).

3. Experiment and Results

3.1. Fabrication of the Proposed Waveguides

The exact reproduction of the designed structure requires tremendous efforts to optimize the parameters of the facility. Fortunately, 3D printing is a process of making prototype parts directly from computer models, which opens up almost unlimited possibilities for rapid prototyping [13]. In this study, waveguides with the gap distance (g) set as 500 µm, 1000 µm, and 2000 µm, respectively, were printed using the SLA 3D printing mechanism Figure 3a shows an overview of the entire 3D printing methodology [25,26]. Figure 3b–e shows a cross-section of the optical electron microscopy (OEM) images of the waveguide (e.g., the gap distance g is 500 µm). The fabricated error of the waveguide structure was measured using a microscope camera to evaluate the accuracy of the dimensions [27] The OEM results proved that the morphology of the 3D-printing was consistent with the design. According to the measurement result, as shown in Figure 3f, d = 800 ± 14.96 µm D = 1500 ± 19 µm, L = 750 ± 18.74 µm, g = 500 ± 14.23 µm. It is noted that the difference between the measured and designed dimensions was within the allowable printer error (<50 µm). Therefore, the fabrication method of the proposed waveguide had the advantages of a low-cost and easy fabrication, and flexible design. These properties are desirable for applications of various terahertz devices [30].

3.2. 3D Printing Based on Stereolithography

The desired waveguide devices were first drafted using commercial CAD drawing software and then translated into STL format, a suitable file for a 3D printer. The file was then sliced in a Z direction using the Photon Workshop software (Version 2.1.23.RC8) The sliced file was then sent to the 3D printer. This SLA printer (ANYCUBIC Photon Mono X) used an inverted lithography set-up with a 405 nm UV light irradiation and LCD screen selective voxel curing that resulted in the finished 3D structure [26]. The transverse resolution was 47 μm and the longitudinal resolution was 1.25 μm (i.e., along the structure height) [23].

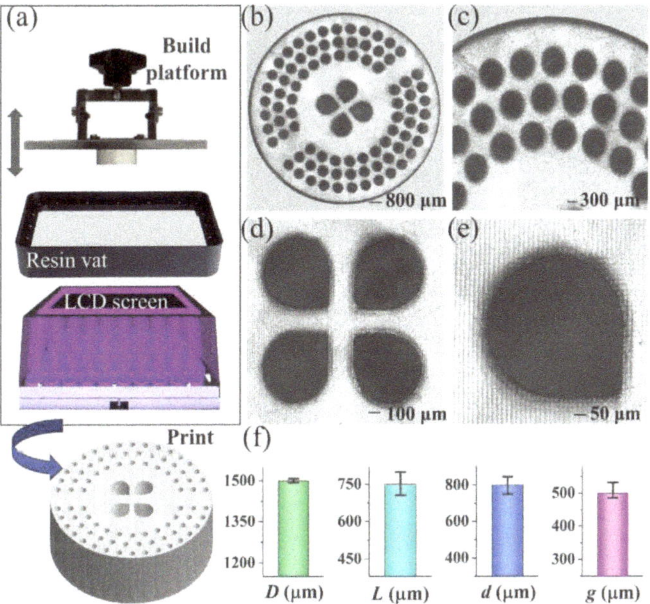

Figure 3. (**a**) Schematic of the SLA printing process and 3D printed object. Optical microscopy images of (**b**) the cross-section of the THz PC waveguide, (**c**) the periodically arranged unit cells of the THz PC waveguide, (**d**) the lilac-shaped resonant cavities of the THz PC waveguide, and (**e**) one petal of the lilac-shaped resonant cavities of the THz PC waveguide. (**f**) Statistical analysis of the size of four parameters after multiple measurements.

3.3. THZ-TDS Setup

All of the experimental investigations were performed with a THz-TDS system (Menlo TeraSmart, Martinsried, Germany). The centered wavelength of the femtosecond laser that was used was 780 nm and the repetition rate was 100 MHz. The frequency resolution of the THz-TDS system was 1.2 GHz with a signal to noise ratio (SNR) of about 80 dB. The beam was guided between the transmitter and detector by off-axis parabolic mirrors.

3.4. Measurement and Discussion

As shown in Figure 4a, in the measurement of reflection loss, the signal is measured by receiver (1#), which is set at the reflection optical path (red arrows). For measurement of the transmission loss, the signal is measured by the receiver (2#) that is set at the transmission optical path (blue arrows). The RI of the photosensitive resin material used in THz frequency range were measured by the THz-TDS system. As shown in Figure 4b,c, the experimental and simulated loss of the proposed waveguide with different gap distances were obtained from 0.1 to 0.5 THz in the potential 6G telecommunication band.

In the reflection spectra, it was found that the resonant frequency (f_R) of the waveguide shifted to the left with an increase in gap distance (g), which is shown in Figure 4b. The experimental and simulated resonant frequency were both in the region of 0.2 to 0.3 THz. The numerical simulation results showed that the reflection loss decreased by about -20 dB, and the experimental results showed that it decreased by about -10 dB. As shown in Figure 4c, in the transmission spectra from 0.1 to 0.3 THz, it was found that the experimental and simulated results demonstrated approximately the same loss, of less than -45 dB, and it showed a remarkable decrease trend higher than 0.3 THz. The magnitudes of loss observed in the experimental results did not exactly match that of the simulated results, and this discrepancy could be attributed to the detection limitation of the experimental apparatus. Overall, the numerical simulation results agreed well with the experimental results, which indicates that the proposed waveguide has a good controllable performance.

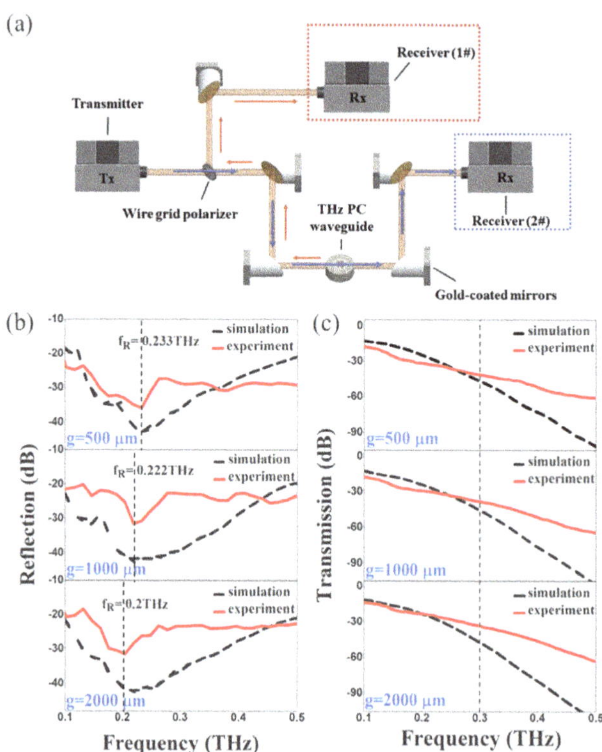

Figure 4. (**a**) Optical path for the THz-TDS measurement setup, the above experiments are performed in dry air. (**b**) The simulated and experimental reflection spectra for the THz PC waveguide with the gap distance (*g*) set as 500 μm, 1000 μm, and 2000 μm, respectively. (**c**) The simulated and experimental transmission spectra for the THz PC waveguide with the gap distance (*g*) set as 500 μm, 1000 μm, and 2000 μm, respectively.

To analyze the resonant performance of the waveguide, the Q factor and reflection loss at the resonant frequency were calculated. Q factor can be expressed by [30]

$$Q = \frac{\omega_r}{\text{FWHM}} \tag{3}$$

where ω_r is the resonant frequency ($\omega_r = 2\pi f_R$) and FWHM is the full width half max of the resonant spectrum. As shown in Figure 5a, the resonant frequency has a shift of 33 GHz. Simultaneously, as shown in Figure 5b, with a gap distance of 500–2000 μm, the measured Q factor was obtained from 1.55 to 2. The reflection loss was further analyzed at the resonant frequency, as shown in Figure 5c. The difference in the reflection loss between the experimental and simulated results was about 10 dB.

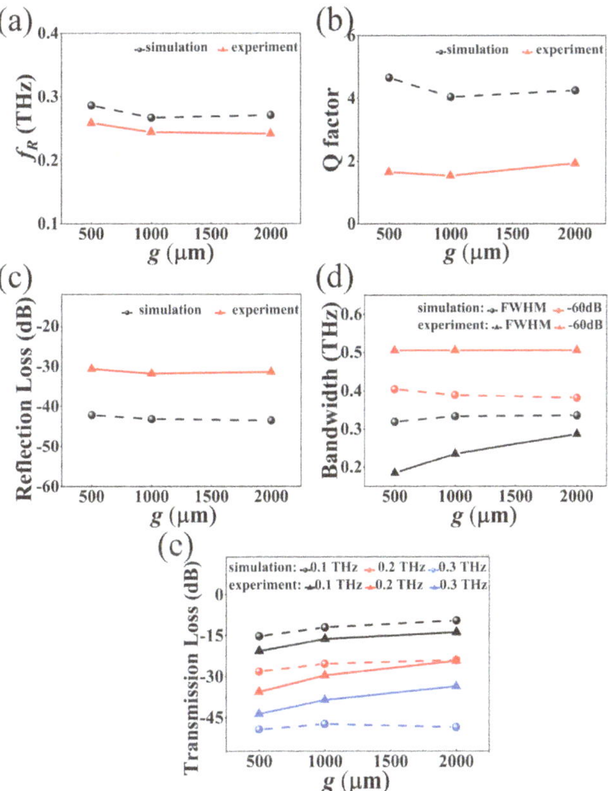

Figure 5. Characteristics of the THz PC waveguide. Influence of the gap distance (g) on (**a**) the resonant frequency f_R, (**b**) Q factor, (**c**) the reflection loss (at $f = f_R$), (**d**) the bandwidth (with FWHM and -60 dB bandwidth, respectively), and (**e**) the transmission loss with $f = 0.1$ THz, $f = 0.2$ THz, and $f = 0.3$ THz, respectively.

In order to better understand the transmission characteristics of the waveguide, the FWHM and the -60 dB bandwidth were analyzed. As shown in Figure 5d, with a gap distance of 500–1000 μm, FWHM was obtained from 0.2 to 0.3 THz. The -60 dB bandwidth was obtained from 0.4 to 0.5 THz, which is within the detection limitation of most commercial THz-TDS systems. In addition, in order to better illustrate the performance of the transmission loss in diverse frequency conditions, the transmission loss was analyzed in detail at different frequencies within 0.3 THz, individually (Figure 5e). The results revealed that the transmission loss decreased with a larger gap distance, and a similar trend was also indicated by the simulation results. The above results show that the waveguide performance could be controlled by the gap distance. Thus, the optimum structural parameters of the waveguide should be selected according to the application requirements.

4. Conclusions

In conclusion, a THz PC waveguide with a lilac-shaped defect is demonstrated in this work. The reflection and transmission characteristics of the proposed waveguide have been analyzed both in simulation and experiment. The waveguide spectral response is further optimized by changing the distance of the lilac-shaped resonant cavities. The designed waveguide performs a resonant reflection in the region of 0.2 to 0.3 THz and a low-pass transmission in the 6G mobile communication window. These results have demonstrated that the waveguide performance can be controlled by the gap distance. For some special occasions, the gap distance can be treated in order to evaluate the pros and cons of this device. In addition, the proposed waveguide has many advantages, including a designable structure, low-cost and easy fabrication, and operation in a low-THz frequency. Although its reflection and transmission loss are larger than other types of waveguides, the performance of the proposed waveguide will be optimized with the development of SLA technology. It holds great promise for extensive applications in communication, sensing, and imaging.

Author Contributions: Conceptualization, J.S. and Y.D.; methodology, J.S., Y.D. and X.L. (Xiuyan Li); software, H.B., X.L. (Xianguo Li), and W.F. (Wei Fan); validation, L.T., W.F. (Weiling Fu), and X.Y.; writing—original draft preparation, J.S. and Y.D.; writing—review and editing, J.S., X.Y. and P.N.; visualization, W.F. (Weiling Fu); supervision, P.N. and J.Y. All authors have read and agreed to the published version of the manuscript.

Funding: This work is sponsored by the National Natural Science Foundation of China (61905177, 61901297, 61903273, and 81802118); the Natural Science Foundation of Tianjin City (19JCQNJC01400); the Open Research Fund of National Mobile Communications Research Laboratory, Southeast University (2022D12); and the Open Fund of IPOC, BUPT (IPOC2021B02).

Institutional Review Board Statement: Not applicable.

Informed Consent Statement: Not applicable.

Data Availability Statement: Not applicable.

Conflicts of Interest: The authors declare no conflict of interest.

References

1. Talataisong, W.; Gorecki, J.; van Putten, L.D.; Ismaeel, R.; Williamson, J.; Addinall, K.; Schwendemann, D.; Beresna, M.; Apostolopoulos, V.; Brambilla, G. Hollow-core antiresonant terahertz fiber-based TOPAS extruded from a 3D printer using a metal 3D printed nozzle. *Photonics Res.* **2021**, *9*, 1513. [CrossRef]
2. Yudasari, N.; Anthony, J.; Leonhardt, R. Terahertz pulse propagation in 3D-printed waveguide with metal wires component. *Opt. Express* **2014**, *22*, 26042–26054. [CrossRef] [PubMed]
3. Lu, Y.; Feng, X.; Wang, Q.; Zhang, X.; Fang, M.; Sha, W.E.I.; Huang, Z.; Xu, Q.; Niu, L.; Chen, X.; et al. Integrated Terahertz Generator-Manipulators Using Epsilon-near-Zero-Hybrid Nonlinear Metasurfaces. *Nano Lett.* **2021**, *21*, 7699–7707. [CrossRef] [PubMed]
4. Li, J.; Zheng, C.; Li, J.; Wang, G.; Liu, J.; Yue, Z.; Hao, X.; Yang, Y.; Li, F.; Tang, T.; et al. Terahertz wavefront shaping with multi-channel polarization conversion based on all-dielectric metasurface. *Photonics Res.* **2021**, *9*, 1939. [CrossRef]
5. Lee, Y.S.; Choi, H.; Kim, B.; Kang, C.; Maeng, I.; Oh, S.J.; Kim, S.; Oh, K. Low-Loss Polytetrafluoroethylene Hexagonal Porous Fiber for Terahertz Pulse Transmission in the 6G Mobile Communication Window. *IEEE Trans. Microw. Theory Tech.* **2021**, *69*, 4623–4630. [CrossRef]
6. Gu, T.; Andryieuski, A.; Hao, Y.; Li, Y.; Hone, J.; Wong, C.W.; Lavrinenko, A.; Low, T.; Heinz, T.F. Photonic and Plasmonic Guided Modes in Graphene–Silicon Photonic Crystals. *ACS Photonics* **2015**, *2*, 1552–1558. [CrossRef]
7. Han, S.; Cong, L.; Srivastava, Y.K.; Qiang, B.; Rybin, M.V.; Kumar, A.; Jain, R.; Lim, W.X.; Achanta, V.G.; Prabhu, S.S.; et al. All-Dielectric Active Terahertz Photonics Driven by Bound States in the Continuum. *Adv. Mater.* **2019**, *31*, e1901921. [CrossRef]
8. Lan, C.; Ma, H.; Wang, M.; Gao, Z.; Liu, K.; Bi, K.; Zhou, J.; Xin, X. Highly Efficient Active All-Dielectric Metasurfaces Based on Hybrid Structures Integrated with Phase-Change Materials: From Terahertz to Optical Ranges. *ACS Appl. Mater. Interfaces* **2019**, *11*, 14229–14238. [CrossRef]
9. Withayachumnankul, W.; Fujita, M.; Nagatsuma, T. Integrated Silicon Photonic Crystals Toward Terahertz Communications. *Adv. Opt. Mater.* **2018**, *6*, 1800401. [CrossRef]
10. Yokoo, A.; Takiguchi, M.; Birowosuto, M.D.; Tateno, K.; Zhang, G.; Kuramochi, E.; Shinya, A.; Taniyama, H.; Notomi, M. Subwavelength Nanowire Lasers on a Silicon Photonic Crystal Operating at Telecom Wavelengths. *ACS Photonics* **2017**, *4*, 355–362. [CrossRef]
11. Reinhard, B.; Torosyan, G.; Beigang, R. Band structure of terahertz metallic photonic crystals with high metal filling factor. *Appl. Phys. Lett.* **2008**, *92*, 201107. [CrossRef]
12. Su, X.; Xu, Q.; Lu, Y.; Zhang, Z.; Xu, Y.; Zhang, X.; Li, Y.; Ouyang, C.; Deng, F.; Liu, Y.; et al. Gradient Index Devices for Terahertz Spoof Surface Plasmon Polaritons. *ACS Photonics* **2020**, *7*, 3305–3312. [CrossRef]
13. Yang, J.; Zhao, J.; Gong, C.; Tian, H.; Sun, L.; Chen, P.; Lin, L.; Liu, W. 3D printed low-loss THz waveguide based on Kagome photonic crystal structure. *Opt. Express* **2016**, *24*, 22454–22460. [CrossRef] [PubMed]
14. Cheben, P.; Halir, R.; Schmid, J.H.; Atwater, H.A.; Smith, D.R. Subwavelength integrated photonics. *Nature* **2018**, *560*, 565–572. [CrossRef] [PubMed]

15. Lin, Y.; Yao, H.; Ju, X.; Chen, Y.; Zhong, S.; Wang, X. Free-standing double-layer terahertz band-pass filters fabricated by femtosecond laser micro-machining. *Opt. Express* **2017**, *25*, 25125–25134. [CrossRef] [PubMed]
16. Headland, D.; Yu, X.; Fujita, M.; Nagatsuma, T. Near-field out-of-plane coupling between terahertz photonic crystal waveguides. *Optica* **2019**, *6*, 1002–1011. [CrossRef]
17. Yee, C.; Jukam, N.; Sherwin, M. Transmission of single mode ultrathin terahertz photonic crystal slabs. *Appl. Phys. Lett.* **2007**, *91*, 194104. [CrossRef]
18. Fan, K.; Xu, X.; Gu, Y.; Dai, Z.; Cheng, X.; Zhou, J.; Jiang, Y.; Fan, T.; Xu, J. Organic DAST Single Crystal Meta-Cavity Resonances at Terahertz Frequencies. *ACS Photonics* **2019**, *6*, 1674–1680. [CrossRef]
19. Gangnaik, A.S.; Georgiev, Y.M.; Holmes, J.D. New Generation Electron Beam Resists: A Review. *Chem. Mater.* **2017**, *29*, 1898–1917. [CrossRef]
20. Huang, Y.; Zhao, Q.; Kalyoncu, S.K.; Torun, R.; Lu, Y.; Capolino, F.; Boyraz, O. Phase-gradient gap-plasmon metasurface based blazed grating for real time dispersive imaging. *Appl. Phys. Lett.* **2014**, *104*, 161106. [CrossRef]
21. Paniagua-Dominguez, R.; Yu, Y.F.; Khaidarov, E.; Choi, S.; Leong, V.; Bakker, R.M.; Liang, X.; Fu, Y.H.; Valuckas, V.; Krivitsky, L.A.; et al. A Metalens with a Near-Unity Numerical Aperture. *Nano Lett.* **2018**, *18*, 2124–2132. [CrossRef] [PubMed]
22. Liu, D.; Zhao, S.; You, B.; Jhuo, S.S.; Lu, J.Y.; Chou, S.; Hattori, T. Tuning transmission properties of 3D printed metal rod arrays by breaking the structural symmetry. *Opt. Express* **2021**, *29*, 538–551. [CrossRef] [PubMed]
23. Ju, X.; Yang, W.; Gao, S.; Li, Q. Direct Writing of Microfluidic Three-Dimensional Photonic Crystal Structures for Terahertz Technology Applications. *ACS Appl. Mater. Interfaces* **2019**, *11*, 41611–41616. [CrossRef] [PubMed]
24. Liu, X.; Tao, J.; Liu, J.; Xu, X.; Zhang, J.; Huang, Y.; Chen, Y.; Zhang, J.; Deng, D.Y.B.; Gou, M.; et al. 3D Printing Enabled Customization of Functional Microgels. *ACS Appl. Mater. Interfaces* **2019**, *11*, 12209–12215. [CrossRef] [PubMed]
25. Ruiz, A.J.; Garg, S.; Streeter, S.S.; Giallorenzi, M.K.; LaRochelle, E.P.M.; Samkoe, K.S.; Pogue, B.W. 3D printing fluorescent material with tunable optical properties. *Sci. Rep.* **2021**, *11*, 17135. [CrossRef]
26. Wang, R.; Yang, W.; Gao, S.; Ju, X.; Zhu, P.; Li, B.; Li, Q. Direct-writing of vanadium dioxide/polydimethylsiloxane three-dimensional photonic crystals with thermally tunable terahertz properties. *J. Mater. Chem. C* **2019**, *7*, 8185–8191. [CrossRef]
27. Yan, D.; Meng, M.; Li, J.; Wang, Y. Proposal for a symmetrical petal core terahertz waveguide for terahertz wave guidance. *J. Phys. D Appl. Phys.* **2020**, *53*, 275101. [CrossRef]
28. Yan, D.; Meng, M.; Li, J.; Wang, Y. Terahertz wave refractive index sensor based on a sunflower-type photonic crystal. *Laser Phys.* **2020**, *30*, 066206. [CrossRef]
29. Ye, L.; Zhang, W.; Ofori-Okai, B.K.; Li, W.; Zhuo, J.; Cai, G.; Liu, Q.H. Super Subwavelength Guiding and Rejecting of Terahertz Spoof SPPs Enabled by Planar Plasmonic Waveguides and Notch Filters Based on Spiral-Shaped Units. *J. Lightwave Technol.* **2018**, *36*, 4988–4994. [CrossRef]
30. Sultana, J.; Islam, M.S.; Cordeiro, C.M.B.; Habib, M.S.; Dinovitser, A.; Kaushik, M.; Ng, B.W.H.; Ebendorff-Heidepriem, H.; Abbott, D. Hollow Core Inhibited Coupled Antiresonant Terahertz Fiber: A Numerical and Experimental Study. *IEEE Trans. Terahertz Sci. Technol.* **2021**, *11*, 245–260. [CrossRef]

Article

Strain Relaxation Effect on the Peak Wavelength of Blue InGaN/GaN Multi-Quantum Well Micro-LEDs

Chaoqiang Zhang [1], Ke Gao [1], Fei Wang [1], Zhiming Chen [1], Philip Shields [2], Sean Lee [3], Yanqin Wang [3], Dongyan Zhang [3], Hongwei Liu [1,*] and Pingjuan Niu [1,*]

[1] Tianjin Key Laboratory of Optoelectronic Detection Technology and System, School of Electronics and Information Engineering, Tiangong University, Tianjin 300387, China; 2031070895@tiangong.edu.cn (C.Z.); 1931075520@tiangong.edu.cn (K.G.); 2010940523@tiangong.edu.cn (F.W.); 1910940502@tiangong.edu.cn (Z.C.)

[2] Department of Electronic and Electrical Engineering, University of Bath, Bath BA2 7AY, UK; p.shields@bath.ac.uk

[3] Sanan Optoelectronics Co., Ltd., Xiamen 361009, China; celee@sanan-e.com (S.L.); tj-wangyanqin@sanan-e.com (Y.W.); dyzhang2012@sanan-e.com (D.Z.)

* Correspondence: liuhongwei@tiangong.edu.cn (H.L.); niupingjuan@tiangong.edu.cn (P.N.)

Abstract: In this paper, the edge strain relaxation of InGaN/GaN MQW micro-pillars is studied. Micro-pillar arrays with a diameter of 3–20 μm were prepared on a blue GaN LED wafer by inductively coupled plasma (ICP) etching. The peak wavelength shift caused by edge strain relaxation was tested using micro-LED pillar array room temperature photoluminescence (PL) spectrum measurements. The results show that there is a nearly 3 nm peak wavelength shift between the micro-pillar arrays, caused by a high range of the strain relaxation region in the small size LED pillar. Furthermore, a 19 μm micro-LED pillar's Raman spectrum was employed to observe the pillar strain relaxation. It was found that the Raman E_2^H mode at the edge of the micro-LED pillar moved to high frequency, which verified an edge strain relaxation of = 0.1%. Then, the exact strain and peak wavelength distribution of the InGaN quantum wells were simulated by the finite element method, which provides effective verification of our PL and Raman strain relaxation analysis. The results and methods in this paper provide good references for the design and analysis of small-size micro-LED devices.

Keywords: InGaN/GaN multiple quantum well (MQW); strain relaxation; micro-LED arrays; photoluminescence (PL); Raman shift

Citation: Zhang, C.; Gao, K.; Wang, F.; Chen, Z.; Shields, P.; Lee, S.; Wang, Y.; Zhang, D.; Liu, H.; Niu, P. Strain Relaxation Effect on the Peak Wavelength of Blue InGaN/GaN Multi-Quantum Well Micro-LEDs. *Appl. Sci.* **2022**, *12*, 7431. https://doi.org/10.3390/app12157431

Academic Editor: Zhi-Ting Ye

Received: 24 June 2022
Accepted: 22 July 2022
Published: 24 July 2022

Publisher's Note: MDPI stays neutral with regard to jurisdictional claims in published maps and institutional affiliations.

1. Introduction

When the size of an LED chip is reduced to tens of microns or even a few microns, it is called a micro-LED chip. Because the micro-display is based on red, green, and blue (RGB) light, micro-LED chips have a high resolution, high brightness, long life, high response speed, and low power consumption. Micro-LEDs have important applications in high-resolution displays, augmented reality, high-speed visible light communication, micro-projectors and other fields [1–3]. Therefore, micro-LED research has been highly valued by researchers in academia and industry all over the world.

In order to improve the luminous efficiency, a multiple quantum well (MQW) structure generally is adopted as the active layer in the micro-LED [4–6]. However, the difference in the lattice constants of the two materials in the quantum well will have a certain impact on the performance of the device. Using the blue LED as an example, stacked InGaN/GaN layers are used to fabricate multiple quantum wells. Due to the ~11% lattice mismatch between InN and GaN when the crystal grows along the *c*-axis, the InGaN/GaN multiple quantum well (MQW) suffers from epitaxial strain and a strong piezoelectric field [7], which leads to a quantum-confined Stark effect (QCSE) that further limits the internal quantum efficiency of InGaN/GaN MQW LEDs [8,9]. In addition, recent studies have indicated that the strain giving rise to the QCSE may be fully or partly relaxed at the boundary of micro-

or nano-scale GaN pillars, stimulating many previous studies on the stress distributions on such samples. For example, Y.Kawakami et al. compared the radiation recombination rate of the quantum well of the micro-column structure at the edge and central regions of the micro-column in a time-resolved spectroscopy test, and the radiation recombination rate of the strain relaxation emission zone was higher than that of the strain zone, verifying the existence of the edge stress release phenomenon [10]. The high-resolution CL test of E.Y.Xie et al. performed a full scan of the sample and observed that there was a difference in the radiation wavelength at the boundary and center of the cylindrical sample [11]. The low photoluminescence measured by Peichen Yu from the embedded InGaN/GaN MQW shows a blueshift energy of 68 meV [12]. The above experiments are conducted with ultra-high-resolution CL and PL spectroscopy, which takes a long time and requires equipment with a very high spatial resolution. Moreover, these studies are all tested on a single MQW pillar and there are no reports studying the edge stress release phenomenon through information obtained from a large number of samples, while the micro-LED applications are generally based on the form of the arrays.

In this paper, several micro-LED arrays of different pillar diameters are fabricated to observe the strain relaxation effect on InGaN/GaN MQWS wavelength modulation. The PL spectra before and after the etching of the micro-LED pillar arrays are used to characterize the MQW edge stress release effect on the arrays' radiation.

Confocal Raman tests are also performed on the single micro-LED pillars to demonstrate the release of lattice mismatch stress on the quantum well sidewalls. Finally, an MQW solid mechanics finite element method (FEM) simulation is used to verify the above analysis.

2. Experiments

2.1. Epitaxial Growth and InGaN/GaN Micro-LED Arrays Fabrication

The InGaN/GaN LED wafer was grown on a 4-inch c-plane sapphire substrate by metal–organic chemical vapor deposition. The epitaxy layer is illustrated in Figure 1: a 3 μm undoped GaN buffer layer; a 1.5 μm n-type GaN layer; twelve periods 3 nm MQWs separated by 12 nm GaN barrier layers; and a 0.2 μm p-type GaN top layer.

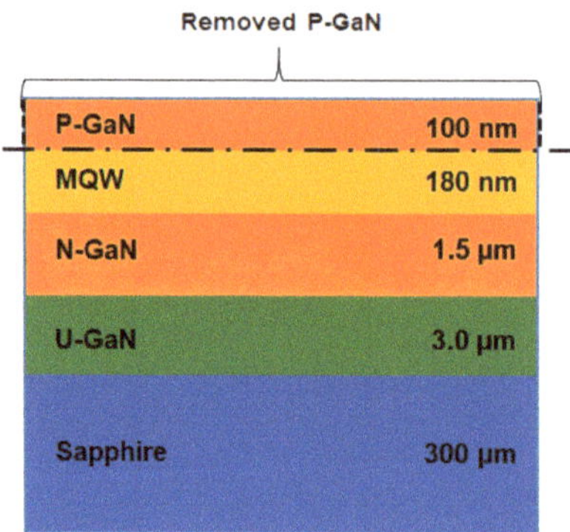

Figure 1. Schematic of GaN blue LED epitaxial wafer layers.

In order to study strain relaxation of InGaN/GaN MQWs, the top p-GaN layer thickness was reduced to 100 nm by inductively coupled plasma (ICP) etching. Then, we divided a 4-inch wafer into 18 areas of the same size with area sizes of 12.24 mm × 19.38 mm. We etched an equal number of micro-pillars of different sizes in each area and the etching period was 30 μm (row: 12,240 μm /30 μm = 408, column: 19,380 μm/30 μm = 646). The diameters of the etched micro-pillars ranged from 3 μm to 20 μm. (as shown in Figure 2)

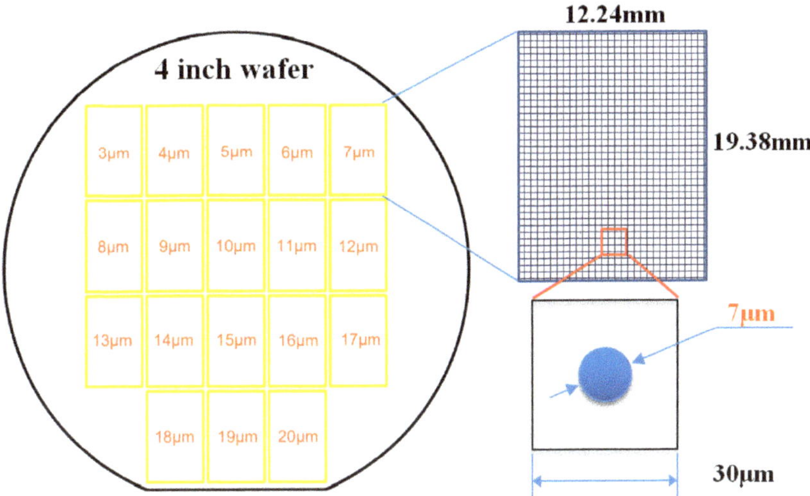

Figure 2. Micro-pillar layout on the 4-inch wafer with diameters from 3 to 20 m.

2.2. PL and Raman Spectrum Measurements of Laser Confocal Scanning Imaging

After the epitaxy wafer was grown, a room-temperature PL mapping was performed with the Etamax photoluminescence tester. The entire epitaxial wafer was scanned by a 40 mW 325 nm laser beam before and after etching. The PL-mapped image resolution is ~1 mm, a total of 7028-pixels.

Then, the single micro-pillars were characterized by room-temperature Raman spectroscopy. Raman measurements were performed using a Horiba XploRA PLUS confocal Raman microscopy system, taking the Z incident direction along the wurtzite GaN *c*-axis and X direction perpendicular to the *c*-axis. The theoretically allowed X polarization $\overline{Z}$ backscattering geometric configuration $Z(X, X)\overline{Z}$ Raman modes for GaN materials are $A_1(LO)$ and E_2^H [13]. The 532 nm Raman laser spot diameter was 0.72 μm and the diffraction grating was set to 2400 grating/mm (high-resolution mode) in order to improve the spectral-spatial resolution.

3. Results and Discussion

3.1. InGaN/GaN Micro-Pillar Arrays PL Spectral Analysis

Figure 3a is the PL peak wavelength distribution of the entire epitaxial wafer after epitaxial growth, which is recorded as the ORI (original wafer PL). The average wavelength of the PL spectrum is 463.04 nm and the standard deviation is 0.82 nm. It can be seen that the emission wavelengths generated by the epitaxial wafer are not completely consistent after epitaxial growth.

Figure 3b shows the distribution of the PL peak wavelength obtained after micro-pillar ICP etching according to the layout shown in Figure 2, denoted as AFT (after etching PL). The obvious PL area outline proves that the etching was successful in dividing the wafer into the 18 smaller areas. The peak wavelength distribution characteristics inherited from the original wafer can also be seen in Figure 3b.

The wavelength distribution inhomogeneity caused by the original wafer epitaxy affects the extraction of micro-pillar PL information. In order to suppress these influences, we subtracted the AFT PL from the ORI PL and filtered the peak wavelength information out of the micro-LED pillar area. Then, some abnormal values far from the average in the ORI minus AFT PL results caused by etching damage and wafer boundary epitaxy quality degradation are filtered based on the Chauvenet-criterion method [14]. The result is shown in Figure 3c and we can obtain the peak wavelength blueshift caused by the micro-LED pillar arrays. There is a wavelength shift of nearly 0–3 nm in the regions of different micro-column sizes. It can be concluded that the smaller the column size, the greater the wavelength blueshift.

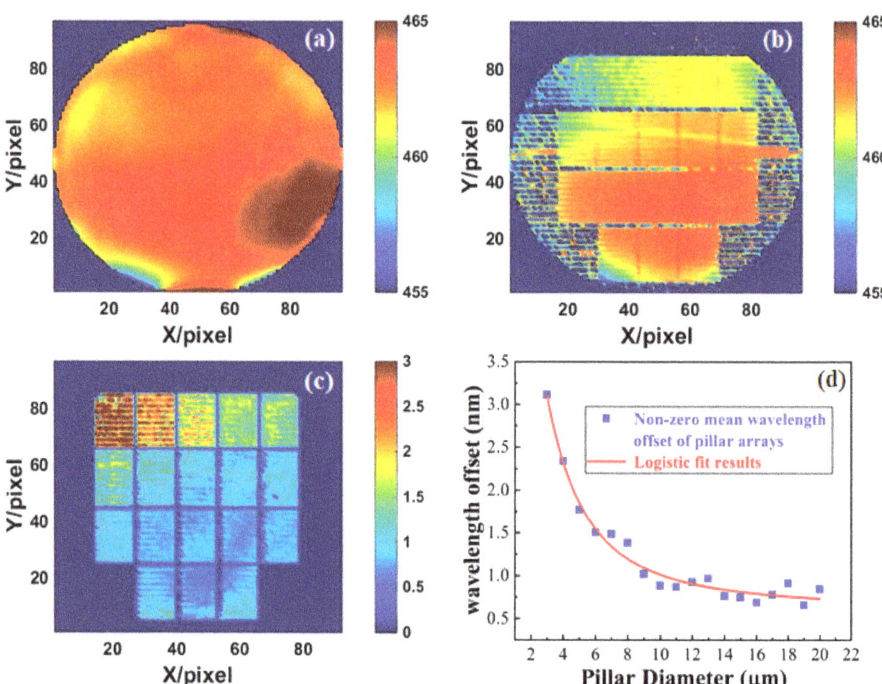

Figure 3. Features of PL spectral data: (**a**) PL spectrum of the GaN blue LED epitaxial wafer before etching, denoted as ORI; (**b**) PL spectrum of the epitaxial wafer after micro-pillar etching, denoted as AFT; (**c**) micro-pillar arrays' wavelength shift, ORI-AFT wavelength shift; (**d**) average wavelength shift for different pillar sizes.

To further illustrate micro-pillar arrays' peak wavelength shift, we employed a non-zero mean method to deal with the peak wavelength of each pillar region. As shown in Figure 3d, the wavelength blueshift gradually decreased from 3.1166 nm in the 3 μm pillar size region to a minimum of 0.6634 nm in the 19 μm pillar size region. We attribute these PL peak wavelength blueshifts to the InGaN/GaN multi-quantum well stress release at the side edge of the micro-pillar. As concluded in previous studies [10–12], the InGaN/GaN quantum well sidewalls' strain relaxation leads to the reduction of the piezoelectric polarization field in the quantum well. Affected by this, the equivalent InGaN/GaN band gap becomes larger and the emission wavelength is blueshifted.

3.2. Single InGaN/GaN Micro-Pillar Raman Measurements and Analysis

In order to verify the above micro-pillar arrays' strain-related PL results, confocal Raman microscopy measurements were employed to characterize micro-pillar strain relaxation. Figure 4 shows the InGaN/GaN wafer Raman spectra collected in the geometric configuration of $Z(X,X)\overline{Z}$. There are distinct peaks at approximately 567 cm^{-1} and 734 cm^{-1}, which are assigned to the E_2^H and A_1^{LO} modes of wurtzite GaN, respectively. The Raman E_2^H mode shift can provide strain relief information due to the biaxial stress caused by lattice mismatch between GaN and InGaN [15,16].

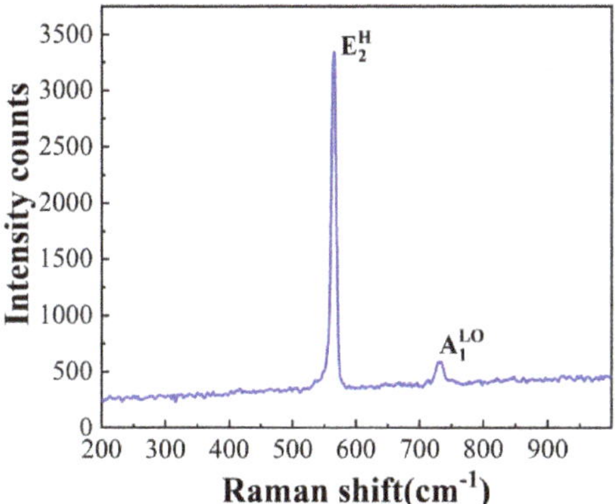

Figure 4. The Raman spectra of InGaN/GaN blue LED wafer.

Through our measurements, the strain relaxation at the edge of the micro-pillar has the same characteristics, so we choose a large InGaN/GaN pillar (diameter of 19 μm) to analyze more Raman information on the pillar surface. The center position of the cylindrical sample was taken as the Raman test starting point, the step size along the radius was approximately 1 μm, and the endpoint was the cylinder boundary. Figure 5a shows the E_2^H mode from the center to the edge and it can be seen that the Raman E_2^H mode tends to shift to the right as the distance increases.

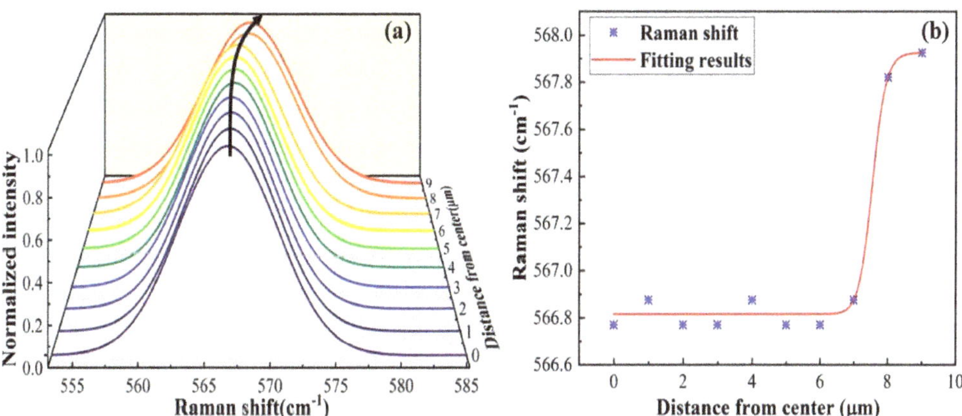

Figure 5. (a) E_2^H modes of Raman spectra after Gaussian fitting and normalization at different distances from center; (b) E_2^H peaks at different distances from the center.

The peak of the E_2^H mode at a different distance from the center was collected in Figure 5b and the movement of the E_2^H peak position could be observed in more detail. As the distance from the center of the micro-pillar increases, the frequency of the Raman peak remains relatively stable near 566.82 cm^{-1} until a distance ≥ 8 μm, when the Raman frequency shift starts to move significantly towards the high-frequency direction. At the boundary position, the frequency shift reaches 567.93 cm^{-1}, which is close to the bulk GaN E_2^H mode. The change of E_2^H indicates that there is strain inside the pillar, while there is full strain relaxation at the edge.

The stress and strain relief at the pillar edge along the *x*-axis ($\Delta\sigma_{xx}$, $\Delta\varepsilon_{xx}$) can be calculated through Equations (1) and (2).

$$\Delta\sigma_{xx} = \frac{\Delta\omega}{k} = \frac{Edge\ Raman\ shif\left(E_2^H\right) - Center\ Raman\ shift\left(E_2^H\right)}{2.56} \tag{1}$$

$$\Delta\varepsilon_{xx} = \frac{\Delta\sigma_{xx}}{C_{11} + C_{12}} \tag{2}$$

Here, k is the stress coefficient, and C_{11} and C_{12} are the elastic moduli of $In_{0.25}Ga_{0.75}N$ (listed in Table 1).

From our fitted results in Figure 5b, the Raman frequency shift near the center area is $566.82 cm^{-1}$ and the Raman frequency shift at the edge position is $567.93\ cm^{-1}$. According to Equations (1) and (2), it can be calculated that there is a stress release of $\Delta\sigma_{xx}$ = 0.43 GPa and a strain release of $\Delta\varepsilon_{xx}$ = 0.10% at the pillar edge.

Furthermore, the micro-pillar strain-related bandgap and peak wavelength shift were derived from Equations (3)–(8). The strain analysis parameters are summarized in Table 1.

Considering a strained InGaN-layer wurtzite crystal pseudomorphically grown along the c-axis on a thick GaN layer, the strain tensor ε in the InGaN well region has the following elements [17]:

$$\varepsilon_{xx} = \varepsilon_{yy} = \frac{a_{GaN} - a_{InGaN}}{a_{InGaN}} \tag{3}$$

$$\varepsilon_{zz} = -2\frac{C_{13}}{C_{33}}\varepsilon_{xx} \tag{4}$$

$$\varepsilon_{xy} = \varepsilon_{xz} = \varepsilon_{yz} = 0 \tag{5}$$

where a_{InGaN} and a_{GaN} are the stress-free lattice parameters of the well and barrier layers, respectively. In this paper, the indium composition in the well layer is 0.25. and the unstressed lattice parameters and elastic constants of $In_{0.25}Ga_{0.75}N$ can be calculated by linear interpolation of the data in [18].

The strain-induced band gap changes were considered, according to the following equations:

$$E_g^{strain} = E_g + (a_{cz} - D_1 - D_3)\varepsilon_{zz} + (a_{ct} - D_2 - D_4)\left(\varepsilon_{xx} + \varepsilon_{yy}\right) \tag{6}$$

where a_{cz} and a_{ct} are the conduction-band deformation potentials along the c-axis and perpendicular to the c-axis, respectively. D_1, D_2, D_3, and D_4 are the shear deformation potentials.

With Equations (4) and (6), the micro-pillar band gap and wavelength shift (ΔE_g^{strain}, $\Delta\lambda$) between the center and edge due to strain relief can be expressed in Equations (7) and (8):

$$\begin{aligned}\Delta E_g^{strain} &= E_g^{strain}(Edge) - E_g^{strain}(Center)\\ &= 2\left(a_{ct} - D_2 - D_4 - \frac{C_{13}}{C_{33}}(a_{cz} - D_1 - D_3)\right)\Delta\varepsilon_{xx}\end{aligned} \tag{7}$$

$$\Delta\lambda = \frac{h \times c}{E_g^{strain}(Edge)} - \frac{h \times c}{E_g^{strain}(Center)} \tag{8}$$

where c is the speed of light in a vacuum and h is Planck's constant.

According to Equations (3)–(8) and our Raman analysis, $\Delta\sigma_{xx}$ and $\Delta\varepsilon_{xx}$, the peak wavelength shift of GaN/InGaN pillar between the center and edge reached 2.45 nm. The Raman shift data show that the strain relaxation at the edge of the pillar is notable, which also provides the reason as to why the small-size InGaN/GaN pillar has a large variation in PL (Figure 3d).

Table 1. Parameters used for GaN and InGaN materials stress and strain analysis [18,19].

Material	InN	GaN	$In_{0.25}Ga_{0.75}N$
k (cm^{-1}/GPa)		2.56	
a (Å)	3.545	3.189	3.278
a_{cz} (eV)	−3.5	−4.9	−4.55
a_{ct} (eV)	−3.5	−11.3	−9.35
D_1 (eV)	−3.7	−3.7	−3.7
D_2 (eV)	4.5	4.5	4.5
D_3 (eV)	8.2	8.2	8.2
D_4 (eV)	−4.1	−4.1	−4.1
C_{11} (Gpa)	223	390	348.25
C_{12} (Gpa)	115	145	137.5
C_{13} (Gpa)	92	106	102.5
C_{33} (Gpa)	224	398	354.5
C_{44} (Gpa)	48	105	90.75

3.3. InGaN/GaN Quantum Well Finite Element Method (FEM) Simulation

From the above section, the strain's relaxation was induced by the Raman shift of the GaN layer, we assume that the stress in the GaN layer is equal to that in the InGaN layer, and the direction is opposite. Then, we use the change of the GaN stress to characterize the stress relaxation in the InGaN quantum well but each InGaN quantum well strain cannot be described exactly.

To support the edge strain relaxation and micro size effects described above, a solid mechanic's finite element model in COMSOL Multiphysics was built to simulate the strain distribution in a single micro-pillar. The isotropy of the in-plane strain in the [0001] growth direction and the cylindrical symmetry simplified the simulation to a two-dimensional case [20]. The simulation model was built according to the structure shown in Figure 1, with the sidewall edge of the pillar set as the start of the x-axis. The free boundary conditions were assigned to the micro-pillar side walls, and the up and bottom GaN surfaces. Meanwhile, we assumed that the quantum wells are pseudo-morphically grown. The initial strain ε_0 between InGaN and GaN layers, calculated by the lattice mismatch Equation (3), was set as the initial strain boundary conditions [21]. The strain and stress were near the quantum well boundary and the carrier radiative recombination was mainly in the quantum wells so that compressive strain of the InGaN quantum wells was observed.

The FEM calculated strain distribution of 12 $In_{0.25}Ga_{0.75}N$ quantum wells is shown in Figure 6a. The height marks the vertical position of the quantum wells in the micro-LED pillar. At the transverse inside of the quantum well (distance from edge > ~600 nm), the in-plane strain ($\varepsilon_{xx} = \varepsilon_{yy}$) of InGaN is about 2.67%, which is the complete strain of $In_{0.25}Ga_{0.75}N$ on GaN. The strain decreases as the boundary approaches. The strain ε_{xx} is about 2.3% near the edge of pillar and the edge strain is too small to represent using a color scale. Therefore, the exact strain distribution is listed in Figure 6b.

Figure 6b shows the calculated average strain of the 12 pairs of QW in a 19 μm diameter pillar. The results show that in the micro-pillars, the strain distribution mainly can be divided into three regions. The first region is the complete strain region, which is located in the middle of the micro-pillars and the strain remains constant at 2.67%. The second region is the gradual strain release region, where the strain decreases from 2.67% to 2.36% within ~600 nm to ~10 nm of the pillar edge. The third area is the mutation region, where strain is released rapidly within 10 nm from the boundary and the strain change to 0.25% at the edge of the pillar.

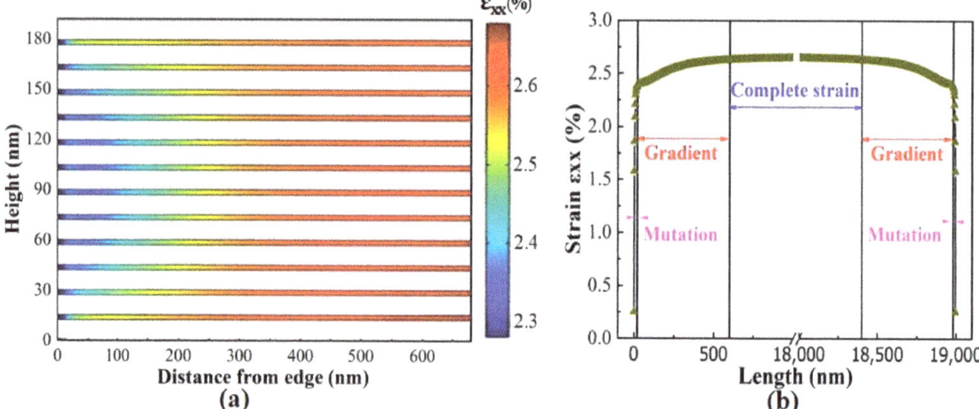

Figure 6. The simulated strain (ε_{xx}) of $In_{0.25}Ga_{0.75}N$ well layer with a diameter of 19 μm GaN/InGaN pillar: (**a**) the strain distribution at the boundaries of the 12 quantum wells; (**b**) the division of strained regions after synthesizing 12 pairs of wells.

Furthermore, based on Equations (3)–(8), the quantum well strain-related PL peak emission was estimated and shown in Figure 7a. In the complete strain region, the quantum well peak wavelength is 459 nm. In the strain mutation region, the strain relaxation makes the peak wavelength move to 416 nm. The emission wavelength in the strained relaxation region is blueshifted by up to 43 nm.

According to the cylindrical symmetry, the distance from the edge (in Figure 7a) is assumed to be the micro-LED pillar radius. With the simulation results in Figure 7a, we statistically analyzed the intensity of the peak wavelengths in the 12 quantum wells in the radius of 100 nm, 200 nm, 400 nm, and 800 nm. The results are shown in Figure 7b. According to our simulations, the 800 nm radius pillar emission peak wavelength is about 457 nm. When the radius is reduced to 400 nm, 200 nm, and 100 nm, the corresponding peak wavelengths are 456 nm, 455 nm, and 451 nm. The peak wavelength offset changes from 1–4 nm. We can draw a conclusion that the piezoelectric polarization generated by the strain in the quantum well directly affects the radiation wavelength. The range of strain relaxation and wavelength shift is mainly related to the thickness of the quantum well and the size of the pillar. The smaller the pillar size, the greater the size modulation effect on the peak wavelength of the device.

With this method, the simulations were expanded to a series of columns with diameters from 3 μm to 20 μm and the peak wavelength shift versus pillar size was explored. As shown in Figure 7c, the red datapoints represent the calculated peak wavelength of GaN/InGaN pillars of different diameter. When the column diameter increases from 3 to 20 um, the peak wavelength shift calculated by the simulation decreases from 1.92 nm to 0.32 nm. With an increase in the diameter of the column, the proportion of the strain relief area decreases continuously and the influence on the peak wavelength of the whole micro-column is weakened.

The black datapoints in Figure 7c represent the results of the corresponding PL experiment (also in Figure 3), which are in good agreement with the trends of the FEM simulation results. However, there is a 0.5–1.1 nm difference between the simulation and the PL test data. We attribute these offset deviations to the following reasons: the P GaN on the original wafer is etched to 100 nm, which will create a strain relaxation from the top of the InGaN/GaN quantum well, and subsequently create a peak wavelength offset increase between the original wafer PL (ORI) and the PL wafer after ICP etching (AFT). On the other hand, the GaN micro-pillar over-etching in the direction of the diameter will create extra edge strain relaxation, which is another reason for the enhancement of the PL peak wavelength offset. These results confirm that the micro size has a modulation effect on the devices' emission wavelength.

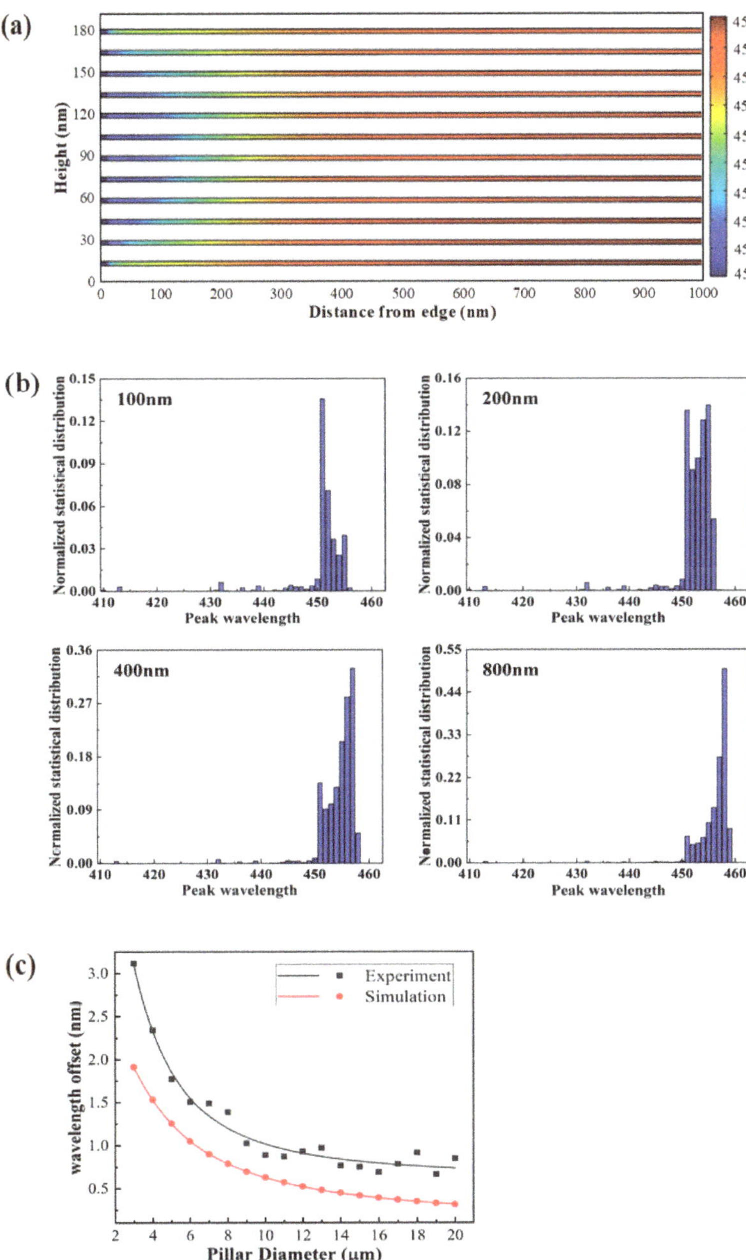

Figure 7. (**a**) The simulated peak wavelength distribution (0–1000 nm) of $In_{0.25}Ga_{0.75}N$ well layer with a diameter of 20 μm; (**b**) the intensity of peak wavelengths in 12 pairs of wells in the radius of 100 nm, 200 nm, 400 nm, 800 nm, starting from the boundary; (**c**) experimental and simulated peak wavelength shift as a function of pillar diameter.

4. Conclusions

In this work, we prepared a 4-inch GaN blue LED wafer by metal–organic chemical vapor deposition (MOCVD). Then, the wafer was divided into 18 areas and each area was fabricated into micro-LED pillar arrays by ICP etching. The pillar diameter of the 18 areas changes from 3 μm to 20 μm. The relationship between the micro-LED pillars' strain relaxation and peak wavelength shift was researched.

Through our PL mapping measurements, the 3 μm micro-LED pillars peak wavelength blueshifts up to 3.1166 nm due to the InGaN/GaN multi-quantum well strain release at the side edge of the micro-pillar. Then, confocal Raman microscopy measurements were performed, and the existence of the pillar edge strain relaxation was verified by E2H mode Raman shift. Finally, the above results were simulated using the finite element method (FEM) in COMSOL software.

This work shows that with the reduction of micro-LED pillar size, the influence on the peak wavelength shift of micro-LEDs increases. The smaller the device size, the more obvious the strain relaxation effect on device peak wavelength. In the future, the high resolution of micro-LED displays could allow the size of pixels and devices to be less than 10 μm. More attention should be paid to the strain relaxation effect on the LED peak wavelength shift. The results of this paper provide important guidance for the design and application of micro-LED devices.

Author Contributions: Conceptualization, H.L. and P.N.; methodology, P.S.; investigation, F.W. and Z.C.; resources, S.L., Y.W. and D.Z.; data curation, K.G.; writing—original draft preparation, C.Z. All authors have read and agreed to the published version of the manuscript.

Funding: This research was funded by the Tianjin Municipal Science and Technology Bureau, Grant 18JCYBJC85400, Grant 18ZXCLGX00090, Grant 20JCQNJC00180 and Grant 19JCTPJC48000; in part by the China Scholarship Council (CSC), Grant 201809345004; and in part by the Tianjin Key Laboratory of Optoelectronic Detection Technology and System under Grant TD13-5035 and Grant 2017ZD06.

Institutional Review Board Statement: Not applicable.

Informed Consent Statement: Not applicable.

Data Availability Statement: Not applicable.

Conflicts of Interest: The authors declare no conflict of interest.

References

1. Wu, Y.; Ma, J.; Su, P.; Zhang, L.; Xia, B. Full-Color Realization of Micro-LED Displays. *Nanomaterials* **2020**, *10*, 2482. [CrossRef] [PubMed]
2. Lee, V.W.; Twu, N.; Kymissis, I. Micro-LED Technologies and Applications. *Inf. Disp.* **2016**, *32*, 16–23. [CrossRef]
3. Tian, P.; Liu, X.; Yi, S.; Huang, Y.; Zhang, S.; Zhou, X.; Hu, L.; Zheng, L.; Liu, R. High-speed underwater optical wireless communication using a blue GaN-based micro-LED. *Opt. Express* **2017**, *25*, 1193–1201. [CrossRef] [PubMed]
4. Zhu, J.; Ning, J.; Zheng, C.; Xu, S.; Zhang, S.; Yang, H. Localized surface optical phonon mode in the InGaN/GaN multiple-quantum-wells nanopillars: Raman spectrum and imaging. *Appl. Phys. Lett.* **2011**, *99*, 113115. [CrossRef]
5. Hsueh, T.-H.; Huang, H.-W.; Kao, C.-C.; Chang, Y.-H.; Ou-Yang, M.-C.; Kuo, H.-C.; Wang, S.-C. Characterization of InGaN/GaN multiple quantum well nanorods fabricated by plasma etching with self-assembled nickel metal nanomasks. *Jpn. J. Appl. Phys* **2005**, *44*, 2661. [CrossRef]
6. Ramesh, V.; Kikuchi, A.; Kishino, K.; Funato, M.; Kawakami, Y. Strain relaxation effect by nanotexturing InGaN/GaN multiple quantum well. *J. Appl. Phys.* **2010**, *107*, 114303. [CrossRef]
7. Wagner, J.; Ramakrishnan, A.; Obloh, H.; Maier, M. Effect of strain and associated piezoelectric fields in InGaN/GaN quantum wells probed by resonant Raman scattering. *Appl. Phys. Lett.* **1999**, *74*, 3863–3865. [CrossRef]
8. Wu, Y.-R.; Chiu, C.; Chang, C.-Y.; Yu, P.; Kuo, H.-C. Size-dependent strain relaxation and optical characteristics of InGaN/GaN nanorod LEDs. *IEEE J. Sel. Top. Quantum Electron.* **2009**, *15*, 1226–1233. [CrossRef]
9. Avit, G.; Robin, Y.; Liao, Y.; Nan, H.; Pristovsek, M.; Amano, H. Strain-induced yellow to blue emission tailoring of axial InGaN/GaN quantum wells in GaN nanorods synthesized by nanoimprint lithography. *Sci. Rep.* **2021**, *11*, 6754. [CrossRef] [PubMed]
10. Kawakami, Y.; Kaneta, A.; Su, L.; Zhu, Y.; Okamoto, K.; Funato, M.; Kikuchi, A.; Kishino, K. Optical properties of InGaN/GaN nanopillars fabricated by postgrowth chemically assisted ion beam etching. *J. Appl. Phys.* **2010**, *107*, 023522. [CrossRef]

11. Xie, E.; Chen, Z.; Edwards, P.; Gong, Z.; Liu, N.; Tao, Y.; Zhang, Y.; Chen, Y.; Watson, I.; Gu, E. Strain relaxation in InGaN/GaN micro-pillars evidenced by high resolution cathodoluminescence hyperspectral imaging. *J. Appl. Phys.* **2012**, *112*, 013107. [CrossRef]

12. Yu, P.; Chiu, C.; Wu, Y.-R.; Yen, H.; Chen, J.; Kao, C.; Yang, H.-W.; Kuo, H.-C.; Lu, T.; Yeh, W. Strain relaxation induced microphotoluminescence characteristics of a single InGaN-based nanopillar fabricated by focused ion beam milling. *Appl. Phys. Lett.* **2008**, *93*, 081110. [CrossRef]

13. Harima, H. Properties of GaN and related compounds studied by means of Raman scattering. *J. Phys. Condens. Matter* **2002**, *14*, R967. [CrossRef]

14. Le Goïc, G.; Brown, C.; Favreliere, H.; Samper, S.; Formosa, F. Outlier filtering: A new method for improving the quality of surface measurements. *Meas. Sci. Technol.* **2012**, *24*, 015001. [CrossRef]

15. Shen, C.; Ng, T.K.; Ooi, B.S.; Cha, D. Strain relief InGaN/GaN MQW micro-pillars for high brightness LEDs. In Proceedings of the 2013 Saudi International Electronics, Communications and Photonics Conference, Riyadh, Saudi Arabia, 27–30 April 2013; pp. 1–3. [CrossRef]

16. Şebnem Çetin, S.; Kemal Öztürk, M.; Özçelik, S.; Özbay, E. Strain analysis of InGaN/GaN multi quantum well LED structures. *Cryst. Res. Technol.* **2012**, *47*, 824–833. [CrossRef]

17. Chuang, S.L. Optical gain of strained wurtzite GaN quantum-well lasers. *IEEE J. Quantum Electron.* **1996**, *32*, 1791–1800. [CrossRef]

18. Vurgaftman, I.; Meyer, J.n. Band parameters for nitrogen-containing semiconductors. *J. Appl. Phys.* **2003**, *94*, 3675–3696. [CrossRef]

19. Zhao, D.; Xu, S.; Xie, M.; Tong, S.; Yang, H. Stress and its effect on optical properties of GaN epilayers grown on Si (111), 6H-SiC (0001), and c-plane sapphire. *Appl. Phys. Lett.* **2003**, *83*, 677–679. [CrossRef]

20. Benabbas, T.; Francois, P.; Androussi, Y.; Lefebvre, A. Stress relaxation in highly strained InAs/GaAs structures as studied by finite element analysis and transmission electron microscopy. *J. Appl. Phys.* **1996**, *80*, 2763–2767. [CrossRef]

21. Christmas, U.M.; Andreev, A.; Faux, D. Calculation of electric field and optical transitions in InGaN/GaN quantum wells. *J. Appl. Phys.* **2005**, *98*, 073522. [CrossRef]

applied sciences

Article

High Repetition Rate, Tunable Mid-Infrared BaGa$_4$Se$_7$ Optical Parametric Oscillator Pumped by a 1 µm Nd:YAG Laser

Yixin He [1], Chao Yan [2,3], Kai Chen [2,3], Degang Xu [2,3,*], Jining Li [2,3], Kai Zhong [2,3], Yuye Wang [2,3], Rui Yao [1], Jiyong Yao [4,*] and Jianquan Yao [2,3]

1   Beijing Institute of Electronic System Engineering, Beijing 100854, China; tju_heyixin@tju.edu.cn (Y.H.); yaorui@casic.com.cn (R.Y.)
2   School of Precision Instruments and Optoelectronics Engineering, Institute of Laser and Optoelectronics, Tianjin University, Tianjin 300072, China; yanchao@tju.edu.cn (C.Y.); chenkai_thz@tju.edu.cn (K.C.); lijining@tju.edu.cn (J.L.); zhongkai@tju.edu.cn (K.Z.); yuyewang@tju.edu.cn (Y.W.); jqyao@tju.edu.cn (J.Y.)
3   Key Laboratory of Optoelectronics Information Technology, Ministry of Education, Tianjin University, Tianjin 300072, China
4   Beijing Center for Crystal Research and Development, Key Lab of Functional Crystals and Laser Technology, Technical Institute of Physics and Chemistry, Chinese Academy of Sciences, Beijing 100190, China
*   Correspondence: xudegang@tju.edu.cn (D.X.); jyao@mail.ipc.ac.cn (J.Y.)

Abstract: A tunable and compact mid-infrared optical parametric oscillator (OPO) based on BaGa$_4$Se$_7$ (BGSe) crystal with a repetition rate up to 250 Hz was demonstrated. A high energy and more approachable side-pumped Q-switched Nd:YAG laser was employed as the pump for the BGSe OPO. Due to the pump double-pass single-resonant oscillator (DP-SRO) configuration, the maximum average power of 250 mW and the maximum pulse energy of 1.28 mJ at 4.06 µm was achieved with the repetition rate of 250 Hz and 100 Hz, respectively. The tunable mid-infrared output from 3.42–4.73 µm was obtained. The influence of repetition rate and cavity length was studied and the thermal effect was analyzed.

Keywords: optical parametric oscillator; BaGa$_4$Se$_7$; mid-infrared

Citation: He, Y.; Yan, C.; Chen, K.; Xu, D.; Li, J.; Zhong, K.; Wang, Y.; Yao, R.; Yao, J.; Yao, J. High Repetition Rate, Tunable Mid-Infrared BaGa$_4$Se$_7$ Optical Parametric Oscillator Pumped by a 1 µm Nd:YAG Laser. *Appl. Sci.* **2022**, *12*, 7197. https://doi.org/10.3390/app12147197

Academic Editor: Edik U. Rafailov

Received: 21 June 2022
Accepted: 14 July 2022
Published: 17 July 2022

Publisher's Note: MDPI stays neutral with regard to jurisdictional claims in published maps and institutional affiliations.

1. Introduction

Widely tunable mid-infrared radiation source operating in 3~5 µm region has been applied to numerous frontier applications, including remote sensing, molecular spectroscopy and atmosphere environmental monitoring [1–3]. OPOs pumped by a Q-switched laser are commonly used in high energy, widely tunable mid-infrared pulse generation. The nonlinear crystal is of great significance to the output performance OPO. Non-oxide crystals usually have a wider transparent range than oxide crystals which have unneglectable multi-phonon absorption over 4 µm [4]. Therefore, non-oxide crystals have advantages in mid-infrared wave generation especially in the range over 4 µm. Due to the defects from the grown process and the limit of the bandgap, the transmittance in visible to near-infrared range of some non-oxide crystals is relatively low. It leads to a limitation on the pump wavelength of non-oxide crystal-based OPOs. Non-oxide crystals used for 1 µm laser pumped OPOs are of great interest, including AgGaS$_2$ [5], HgGa$_2$S$_4$ [6], LiInSe$_2$ [7], LiGaS$_2$ [8], BaGa$_4$S$_7$ [9], and CdSiP$_2$ [10]. High-quality non-oxide crystals which can be used for a commercial 1 µm laser pumped OPO are still urgently needed.

The newly developed mid-infrared crystal BaGa$_4$Se$_7$ (BGSe) exhibits a wide bandgap (2.64 eV), wide transparent range (0.47–18 µm) and high laser damage threshold (557MW/cm^2), which is beneficial for OPOs pumped by an economical 1064 nm laser [11]. Generation of high energy mid-infrared laser has been reported in both 3~5 µm mid-wave mid-infrared [12–14] and 8~14 µm long-wave mid-infrared range [14–16]. The maximum energy of 21.5 mJ/pulse has been reported using a large-size BGSe of 10 × 10 × 16 mm^3 [17]. However, due to the relatively low thermal conductivity [18] and the increasing damage probability under high repetition rate laser pumping [19–21], reports on the high average power BGSe-OPO pumped with 1064 nm laser are rare.

In this paper, we demonstrated a widely tunable BGSe optical parametric oscillator with a repetition rate up to 250 Hz. Pumped by a side-pumped Q-switched Nd:YAG laser, the BGSe-OPO reached a maximum output average power of 250 mW at 4.06 μm and an optical-to-optical conversion efficiency of 11.1%, which is the highest in the current results regarding 1064 nm laser pumped high repetition rate mid-infrared source based on BGSe crystal. Using a pump double-pass configuration, the OPO threshold was reduced to 2.65 mJ/pulse with a cavity length of 30 mm. Tunable mid-infrared output from 3.42 μm to 4.73 μm is achieved by changing the critical phase-matching condition. The pulse width of the generated mid-infrared wave was measured to be 13.6 ns and the linewidth at 4061 nm is estimated to be 3.92 nm.

2. Experimental Setup

The experimental setup of the BGSe-based OPO is shown in Figure 1. A side-pumped electro-optic Q-switched Nd:YAG laser was adopted as the fundamental 1064 nm pump source. The side pump module was composed of five quasi-CW 808 nm laser diode (LD) arrays and a 148 mm long Nd:YAG rod (0.6% doped). The pulse width of the quasi-CW LD arrays was 200 μs and the repetition rate was adjustable in the range of 1–1000 Hz. The electro-optic Pockels cell consisted of two BBO crystals with dimensions of 6 mm × 6 mm × 20 mm. The laser cavity was a plano-plano cavity with a high reflection mirror M1, an output coupler M2, a quarter wave plate and a Brewster polarizer. The output coupler M2 was partial reflection (R ≈ 5%) to avoid laser damage to the crystal rod caused by the high energy intensity in the cavity. The total length of the laser cavity was 350 mm.

A half wave plate and a Brewster polarizer were combined as an energy attenuator to adjust the pump energy and maintained the polarization direction. An optical isolator was employed to avoid the possible reflected damage to the Nd:YAG laser. The DP-SRO BGSe-OPO was composed of two BaF$_2$ mirror M3 (1.064 μm AR and 1.35–1.65 μm and 3.65–4.50 μm HR) and M4 (3.65–4.50 μm AR and 1.064 μm and 1.35–1.65 μm HR). The DP-SRO configuration can effectively improve the mid-infrared output and reduce the threshold [22].

A large-size BGSe crystal (shown in the inset of Figure 1) with dimensions of 8 mm × 8 mm × 15 mm was used as the nonlinear optical crystal in OPO. The BGSe crystal employed was grown by vertical Bridgman method [23]. In addition, two zone annealing technique was used to improve the optical properties. The BGSe crystal exhibits high transmittance from 0.47 to 18 μm and low absorption coefficient of 0.042 cm^{-1} at 4 μm. The BGSe crystal was cut at θ = 42.5° and φ = 0° for Type I phase-matching condition (o-ee). Both sides of the BGSe crystal were well-polished and AR-coated for 1.064 μm, 1.3–1.6 μm and 3–5 μm. The crystal was wrapped in indium foil and cooled to 25 °C with brass heat sink to reduce the thermal effect in BGSe crystal. The wavelength of the signal wave was measured by a spectrometer (Agilent, 86142B, Santa Clara, CA, USA), which was used to calculated the wavelength of the generated mid-infrared wave. A dichroic BaF$_2$ mirror (3.65–4.50 μm AR and 1.064 μm and 1.35–1.65 μm HR) was used as the long-wave pass filter. The energy of the generated mid-infrared wave was measured by a mid-infrared energy detector (Newport corporation, 919E-0.1-12-25K, Irvine, CA, USA).

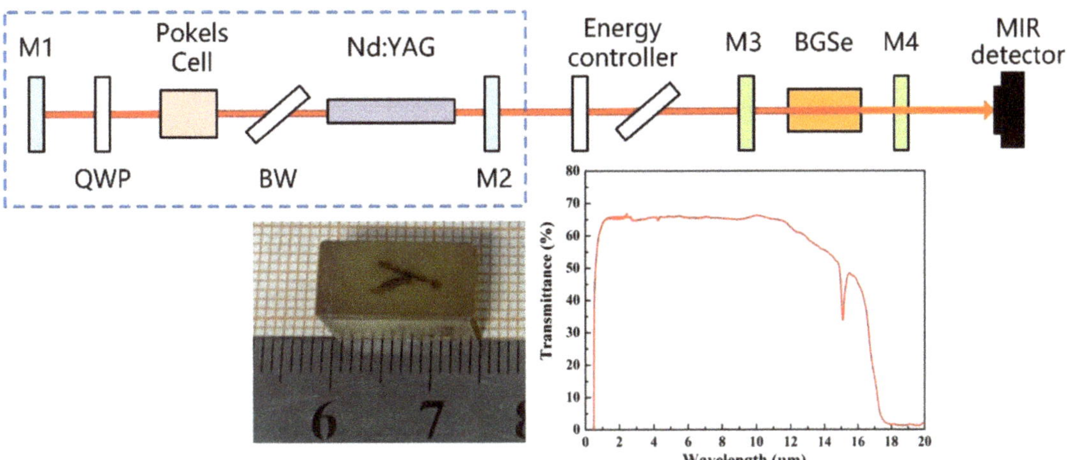

Figure 1. Schematic of BaGa$_4$Se$_7$-OPO. The inset shows the photograph of the BaGa$_4$Se$_7$ crystal and the transmittance curve (Adapted with permission from Ref. [23]. Copyright © 2019, American Chemical Society).

3. Results and Discussion

The input-output characteristics in Q-switched operation are shown in Figure 2. The maximum output energies of 22.0 mJ/pulse, 19.1 mJ/pulse, 17.7 mJ/pulse and 15.1 mJ/pulse were obtained with no saturation when the repetition rates were 100 Hz, 150 Hz, 200 Hz and 250 Hz, respectively. According to the theoretical analysis of transient thermal distribution of repetitively pumped laser by W. Koechner, the temperature in the center of the laser rod increases with the increase of the repetition rate [24]. Serious heat accumulation at higher repetition rate will broaden the fluorescent spectrum and shorten the upper-level lifetime, leading to the decrease of laser output. The higher pump current was not tested to avoid the damage to the Nd:YAG crystal. The lasing thresholds were about 45 A for different repetition rates. The spatial profile of pump beam at the front surface of BGSe crystal was measured as shown in the inset of Figure 2 (linear in spatial dimensions and color scale). The 1064 nm pump beam diameter was 2.5 mm (vertical direction) × 1.0 mm (horizontal direction). The spot size in horizontal direction was shorter than in vertical direction due to the aberration caused by the Brewster polarizer.

The input-output characteristics of the generated mid-infrared wave at 4.06 μm with different repetition rates are shown in Figure 3a. With the increase of repetition rate, the LIDT decreased rapidly. When the repetition rate was above 150 Hz, damage on the surface of the BGSe crystal was observed with pump energy higher than 9 mJ/pulse (corresponding to peak intensity of about 20 MW/cm^2). For higher repetition rate of 500–1000 Hz, optical damages were much more intense, which limited the further improvement in the repetition rate of the BGSe-OPO. The LIDT is lower than our recent report on the low repetition rate (10 Hz) experiment [25]. The thermal effect caused by the extra crystal absorption to the pump wave and the relatively low beam quality from the side-pumped laser were other possible reasons.

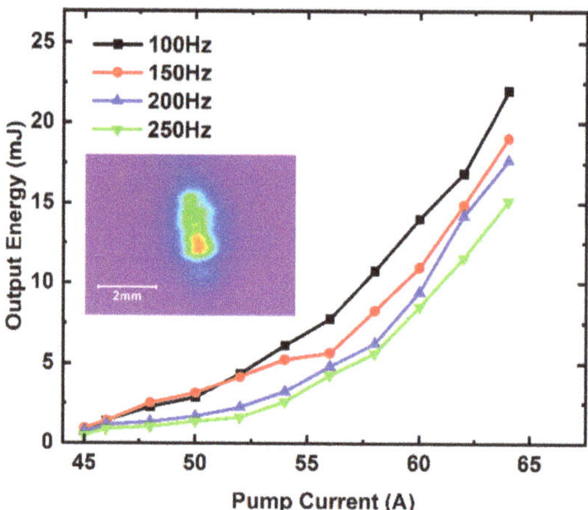

Figure 2. Input-output characteristics of side-pumped Nd:YAG laser. Inset: spatial profile of pump beam at the front surface of BGSe crystal.

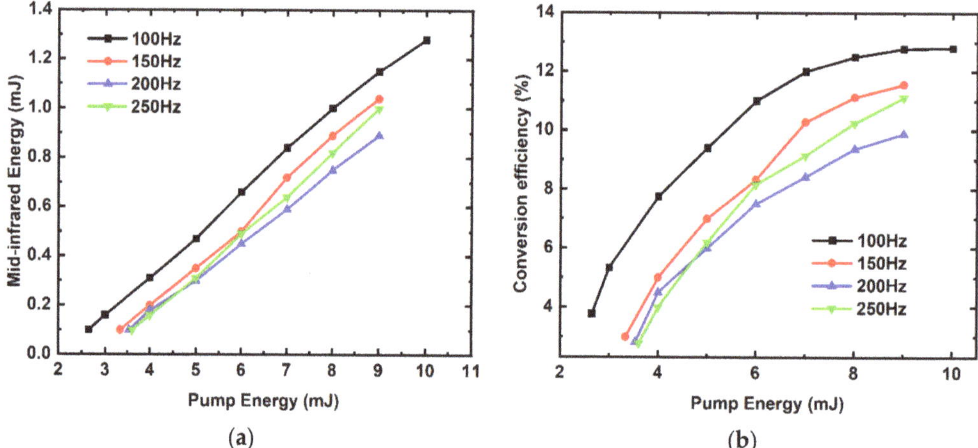

Figure 3. (**a**) The input-output characteristics of the BGSe-based DP-SRO with different repetition rate; (**b**) The conversion efficiency of the BGSe-based DP-SRO with different repetition rate.

The maximum output energy of 1.28 mJ/pulse was achieved at pump energy of 10 mJ/pulse with a repetition rate of 100 Hz. The maximum output average power was 250 mW with a repetition rate of 250 Hz. A slight increase in the OPO threshold was observed with the increase of the repetition rate. The OPO thresholds were measured to be 2.65 mJ/pulse, 3.34 mJ/pulse, 3.54 mJ/pulse and 3.61 mJ/pulse for the repetition rate of 100 Hz, 150 Hz, 200 Hz and 250 Hz, respectively.

Figure 3b shows the conversion efficiencies of the BGSe-OPO with different repetition rates. The maximum conversion efficiency of 12.8% was obtained for the repetition rate of 100 Hz. For different repetition rates of 100 Hz, 150 Hz, 200 Hz and 250 Hz, the slope efficiencies were measured to be 16.4%, 17.0%, 14.4% and 16.6%, respectively. No saturation phenomenon was observed because the low pump peak intensity was limited by the LIDT. Further optimization to pump beam quality could improve the output characteristics of the BGSe-OPO.

The output energy and conversion efficiency of BGSe-OPO decreased with the increase of repetition rate due to the severe thermal effect. It is especially significant for the mid-infrared crystal with poor thermal conductivity like $AgGaS_2$ ($1.4\,\mathrm{W\,m^{-1}\,K^{-1}}$), $AgGaSe_2$ ($1.0\,\mathrm{W\,m^{-1}\,K^{-1}}$) and $BaGa_4Se_7$ ($0.56\,\mathrm{W\,m^{-1}\,K^{-1}}$) [11,18]. With low thermal conductivity, the heat absorbed by the crystal from the pump pulse would accumulate because the pulse interval is shorter than the thermal-relaxation time. With the repetition rate increasing, the thermal effect would be more harmful [26]. The thermal diffusivity is $0.502\,\mathrm{mm^2\,s^{-1}}$ for BGSe crystal along the *a* crystallographic axis [18], while the thermal diffusion time constant τ is about 376 ms, which is much longer than commonly used mid-infrared crystal, e.g., 5.7 ms for ZGP crystal. The prolonged accumulation of heat would lead to thermal lens effect and thermal dephasing which is unbeneficial to the parametric progress. Improving the cooling structure and a using narrow pump beam with high quality could rapidly reduce the thermal effects in BGSe crystal. The increase of the output energy of OPO at 250 Hz may be the joint result of the thermal focusing effect of Nd:YAG and the stabilization of the thermal effect in BGSe crystal.

The influence of the OPO cavity length to the input-output characteristics were studied at 4.06 μm with the fixed repetition rate at 100 Hz, as shown in Figure 4a. For different cavity lengths of 30 mm, 50 mm and 80 mm, the maximum output energies were 1.28 mJ/pulse, 1 mJ/pulse and 0.87 mJ/pulse, respectively. With the decrease of the cavity length, the transit period of the signal wave in the OPO cavity would decrease, which was beneficial to the interaction between the pump and the signal wave and led to the increase of the mid-infrared output energy.

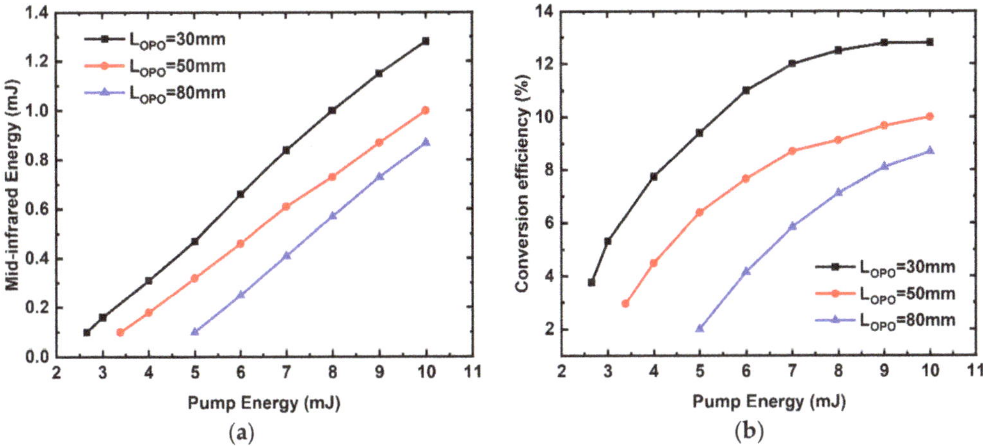

Figure 4. (**a**) The input-output characteristics of the BGSe-based DP-SRO with different cavity lengths; (**b**) The conversion efficiency of the BGSe-based DP-SRO with different cavity lengths.

The OPO thresholds were measured to be 2.65 mJ/pulse, 3.38 mJ/pulse and 5.00 mJ/pulse for the cavity lengths of 30 mm, 50 mm and 80 mm, respectively. The experimental results verify the OPO theory established by S. J. Brosnan and R. L. Byer [22]. With the decrease of the cavity length, the instantaneous cavity net gain increased which leads to the decrease of threshold and the increase of conversion efficiency. The conversion efficiencies of the BGSe-OPO with different cavity lengths were shown in Figure 4b. The maximum conversion efficiency of 12.8% was obtained for the cavity length of 30 mm. For different cavity lengths of 30 mm, 50 mm and 80 mm, the slope efficiencies were measured to be 16.4%, 13.7% and 15.6%. The slope efficiency was measured to be almost constant for each cavity length, which means the back-conversion effect is negligible. With the increase of pump energy, the thermal effect caused by crystal absorption to the pump wave became more serious, causing the decrease of conversion efficiency.

By rotating the BGSe crystal, the tuning range of 3.62–4.10 µm was obtained in the BGSe-OPO. The theoretical wavelength-tuning curves versus the phase-matching angle of Type I BGSe-OPO is calculated based on the Sellmeier equations given in [27]:

$$n_x^2 = 5.952953 + \frac{0.250172}{\lambda^2 - 0.081614} - 0.001709 \cdot \lambda^2 \tag{1}$$

$$n_y^2 = 6.021794 + \frac{0.256951}{\lambda^2 - 0.079191} - 0.001925 \cdot \lambda^2 \tag{2}$$

$$n_z^2 = 6.293976 + \frac{0.282648}{\lambda^2 - 0.094057} - 0.002579 \cdot \lambda^2 \tag{3}$$

The experimental results fit well with the theoretical calculation in Figure 5a. Figure 5b shows the tuning output characteristics of the BGSe-OPO. The tunable output decreased at both sides of the tuning curve due to the increasing pump loss from the Fresnel reflection. The abnormal valley of the tuning curve in Figure 5b around 4.34 µm was deduced to be related to the crystal impurities and defects.

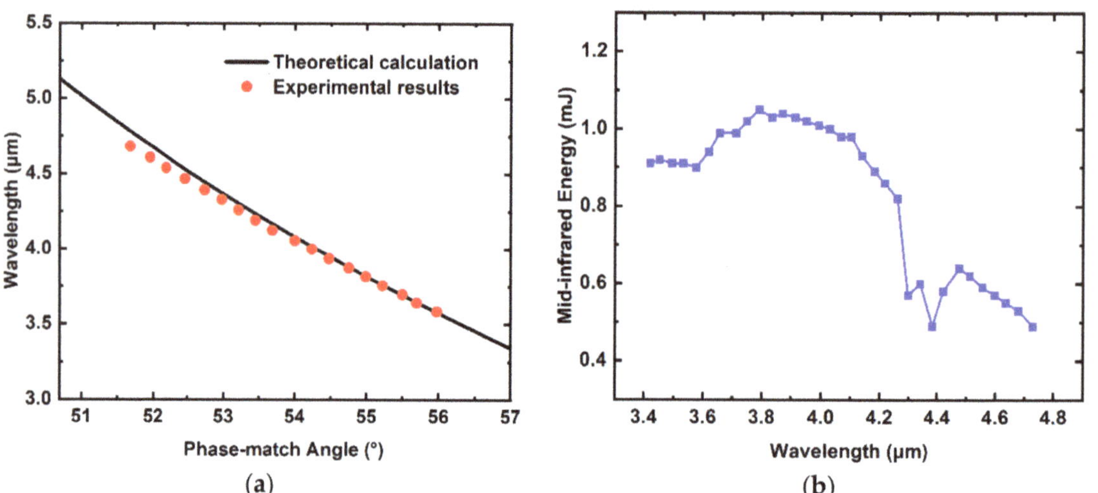

Figure 5. (**a**) Comparison of the tuning curve between the experimental and the calculated results. (**b**) The tuning output characteristics of the BGSe-OPO.

Figure 6 shows the temporal pulse profiles of the pump wave and the mid-infrared wave, measured by Si photodiode (Thorlabs, DET025A/M, Newton, NJ, USA) and (HgCdZn) Te photodiode (Vigo, PCI-9, Ozarow, Mazowiecki, Poland), respectively. With the rapid decline of pump intensity at the trailing edge, the pulse width of the mid-infrared wave (13.6 ns) was measured to be shorter than the pump wave (20.0 ns) [28].

The spectrum of signal wave at normal incidence was recorded by a spectrometer, as shown in Figure 7. The Gaussian fitted curve has a full width at half maxima (FWHM) of 0.49 nm with a central wavelength of 1442.42 nm. According to the phase-match condition, the FWHM of generated mid-infrared is estimated to be 3.92 nm at 4061 nm, much narrower than the 2 µm pumped degenerate parametric conversion [21].

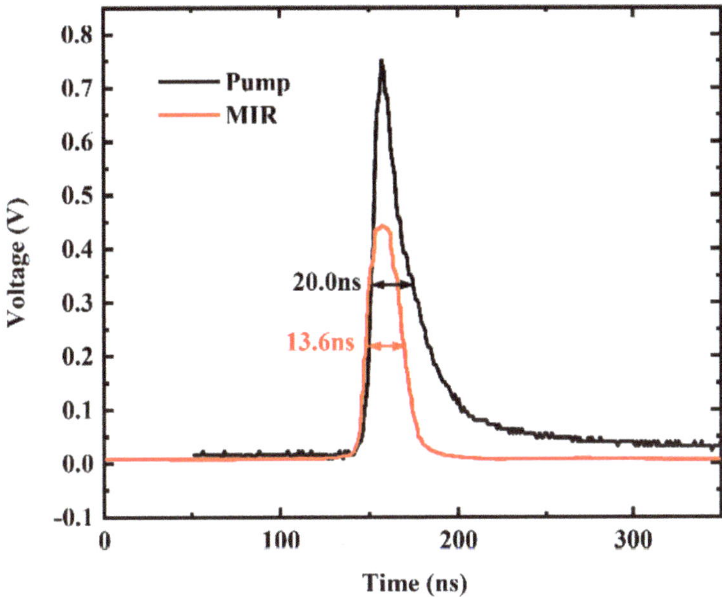

Figure 6. Comparison of the temporal pulse profiles of the pump wave and the mid–infrared wave.

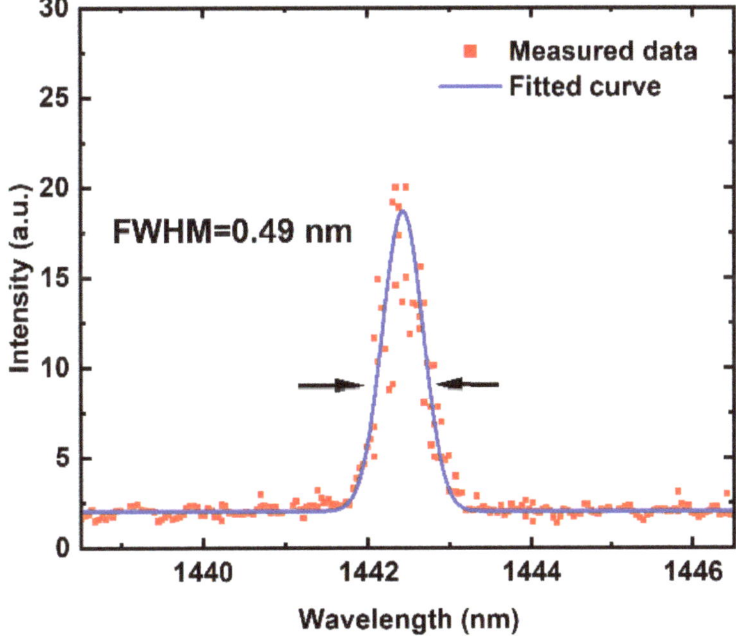

Figure 7. The signal wave spectrum at normal incidence.

4. Discussion and Conclusions

In this work, a 1064 nm laser pumping, tunable BGSe-OPO was demonstrated. Using a side-pumped electro-optical Q-switched Nd:YAG laser as the fundamental pump source, a tunable mid-infrared wave with high repetition rate up to 250 Hz was generated from

BGSe-OPO. DP-SRO configuration was utilized to improve the conversion efficiency and reduce the OPO threshold. The maximum average power of 250 mW was achieved at 4.06 μm, with the OPO cavity length of 30 mm. The influences of the repetition rate and the cavity length were studied experimentally. By rotating the BGSe crystal, tunable mid-infrared output from 3.42–4.73 μm was achieved. The pulse width of the generated mid-infrared wave was measured to be 13.6 ns and the linewidth at 4061 nm was estimated to be 3.92 nm.

The thermal effect of BGSe crystal was observed to be significant, especially in the high repetition rate OPO. Additionally, the accumulation of the thermal effect would result in the decrease of output power and LIDT with the increase of repetition rate, which was harmful to the parametric conversion process. Further optimizations in the pump beam quality, cooling method and cavity structure of DP-OPO would improve the output characteristics of BGSe-OPO.

The BGSe-OPO has advantages in generating the mid-infrared wave with widely tuning range, high power and high repetition rate. Furthermore, BGSe-OPO can be pumped with a 1064 nm laser which is low-cost and usually has a simple structure. BGSe-OPO has the advantages of both the tuning range of ZGP-OPO and the simple structure of PPLN-OPO, which are commonly used in commercial products. The system in this work will show great commercialization potential with the utilization of a commercial 1064 nm laser with a compact size.

Author Contributions: Conceptualization, R.Y. and J.Y. (Jiyong Yao); methodology, D.X.; validation, C.Y., K.C. and Y.H.; formal analysis, Y.W.; investigation, K.Z.; resources, J.Y. (Jiyong Yao); data curation, C.Y.; writing—original draft preparation, K.C.; writing—review and editing, Y.H.; visualization, K.C.; supervision, J.L.; project administration, D.X.; funding acquisition, D.X. and J.Y. (Jianquan Yao). All authors have read and agreed to the published version of the manuscript.

Funding: This research was funded by National Natural Science Foundation of China (NSFC), grant numbers U1837202, 62175182 and 62011540006.

Institutional Review Board Statement: Not applicable.

Informed Consent Statement: Not applicable.

Data Availability Statement: Not applicable.

Conflicts of Interest: The authors declare no conflict of interest.

References

1. Sigrist, M. Mid-infrared laser-spectroscopic sensing of chemical species. *J. Adv. Res.* **2015**, *6*, 529–533. [CrossRef] [PubMed]
2. Quagliano, J.; Stoutland, P.; Petrin, R.; Sander, R.; Romero, R.; Whitehead, M.; Quick, C.; Tiee, J.; Jolin, L. Quantitative chemical identification of four gases in remote infrared (9–11 μm) differential absorption lidar experiments. *Appl. Opt.* **1997**, *36*, 1915–1927. [CrossRef] [PubMed]
3. Das, S. Broadly tunable multi-output coherent source based on optical parametric oscillator. *Opt. Laser Technol.* **2015**, *71*, 63–67. [CrossRef]
4. Schunemann, P.; Zawilski, K.; Pomeranz, L.; Creeden, D.; Budni, P. Advances in nonlinear optical crystals for mid-infrared coherent sources. *J. Opt. Soc. Am. B* **2016**, *33*, D36–D43. [CrossRef]
5. Fan, Y.; Eckardt, R.; Byer, R.; Route, R.; Feigelson, R. AgGaS$_2$ infrared parametric oscillator. *Appl. Phys. Lett.* **1984**, *45*, 313–315. [CrossRef]
6. Badikov, V. A HgGa$_2$S$_4$ optical parametric oscillator. *Quantum Electon.* **2003**, *33*, 831–832. [CrossRef]
7. Zondy, J.; Vedenyapin, V.; Yelisseyev, A.; Lobanov, S.; Isaenko, L.; Petrov, P. LiInSe$_2$ nanosecond optical parametric oscillator. *Opt. Lett.* **2005**, *30*, 2460–2462. [CrossRef]
8. Tyazhev, A.; Vedenyapin, V.; Marchev, G.; Isaenko, L.; Kolker, D.; Lobanov, S.; Petrov, P.; Yelisseyev, A.; Starikova, M.; Zondy, J. Singly-resonant optical parametric oscillation based on the wide band-gap mid-IR nonlinear optical crystal LiGaS$_2$. *Opt. Mater.* **2013**, *35*, 1612–1615. [CrossRef]
9. Tyazhev, A.; Kolker, D.; Marchev, G.; Badikov, V.; Badikov, D.; Shevyrdyaeva, G.; Panyutin, V.; Petrov, V. Midinfrared optical parametric oscillator based on the wide-bandgap BaGa$_4$S$_7$ nonlinear crystal. *Opt. Lett.* **2012**, *37*, 4146–4148. [CrossRef]
10. Petrov, V.; Schunemann, P.; Zawilski, K.; Pollak, T. Non critical singly resonant optical parametric oscillator operation near 6.2 microm based on a CdSiP$_2$ crystal pumped at 1064 nm. *Opt. Lett.* **2009**, *34*, 2399–2401. [CrossRef]

11. Yao, J.; Mei, D.; Bai, L.; Lin, Z.; Yin, W.; Fu, P.; Wu, Y. BaGa$_4$Se$_7$: A New Congruent-Melting IR Nonlinear Optical Material. *Inorg. Chem.* **2010**, *49*, 9212–9216. [CrossRef]
12. Yuan, J.-H.; Li, C.; Yao, B.-Q.; Yao, J.-Y.; Duan, X.-M.; Li, Y.-Y.; Shen, Y.-J.; Wu, Y.-C.; Cui, Z.; Dai, T.-Y. High power, tunable mid-infrared BaGa$_4$Se$_7$ optical parametric oscillator pumped by a 2.1 μm Ho: YAG laser. *Opt. Express* **2016**, *24*, 6083–6087. [CrossRef]
13. Yang, F.; Yao, J.; Xu, H.; Feng, Y.; Yin, W.; Li, F.; Yang, J.; Du, S.; Peng, Q.; Zhang, J.; et al. High efficiency and high peak power picosecond mid-infrared optical parametric amplifier based on BaGa$_4$Se$_7$ crystal. *Opt. Lett.* **2013**, *38*, 3903–3905. [CrossRef]
14. Kostyukova, N.; Boyko, A.; Badikov, V.; Badikov, D.; Shevyrdyaeva, G.; Panyutin, V.; Marchev, G.; Kolker, D.; Petrov, V. Widely tunable in the mid-IR BaGa$_4$Se$_7$ optical parametric oscillator pumped at 1064 nm. *Opt. Lett.* **2016**, *41*, 3667–3670. [CrossRef]
15. Xu, D.; Zhang, J.; He, Y.; Wang, Y.; Yao, J.; Guo, Y.; Yan, C.; Tang, L.; Li, J.; Zhong, K.; et al. High-energy, tunable, long-wave mid-infrared optical parametric oscillator based on BaGa$_4$Se$_7$ crystal. *Opt. Lett.* **2020**, *45*, 5287–5290. [CrossRef]
16. Yang, F.; Yao, J.; Guo, Y.; Yuan, L.; Bo, Y.; Peng, Q.; Cui, D.; Yu, Y.; Xu, Z. High-energy continuously tunable 8–14 μm picosecond coherent radiation generation from BGSe-OPA pumped by 1064 nm laser. *Opt. Laser Technol.* **2020**, *125*, 106040. [CrossRef]
17. Zhang, Y.; Zuo, Y.; Li, Z.; Wu, B.; Yao, J.; Shen, Y. High energy mid-infrared laser pulse output from a BaGa$_4$Se$_7$ crystal-based optical parametric oscillator. *Opt. Lett.* **2020**, *45*, 4595–4598. [CrossRef]
18. Yao, J.; Yin, W.; Feng, K.; Li, X.; Mei, D.; Lu, Q.; Ni, Y.; Zhang, Z.; Hu, Z.; Wu, Y. Growth and characterization of BaGa$_4$Se$_7$ crystal. *J. Cryst. Growth* **2012**, *346*, 1–4. [CrossRef]
19. Kostyukova, N.; Boyko, A.; Erushin, E.; Kostyukov, A.; Badikov, V.; Badikov, D.; Kolker, D. Laser-induced damage threshold of BaGa$_4$Se$_7$ and BaGa$_2$GeSe$_6$ nonlinear crystals at 1.053 μm. *J. Opt. Soc. Am. B* **2020**, *37*, 2655–2659. [CrossRef]
20. Yang, K.; Liu, G.; Li, C.; Yao, B.; Yao, J.; Chen, Y.; Mi, S.; Duan, X.; Dai, T. Research on performance improvement technology of a BaGa$_4$Se$_7$ mid-infrared optical parametric oscillator. *Opt. Lett.* **2020**, *45*, 6418–6421. [CrossRef]
21. Liu, G.; Chen, Y.; Li, Z.; Yang, K.; Yao, B.; Yao, J.; Wang, R.; Yang, C.; Mi, S.; Dai, T.; et al. High-beam quality ? 1 μm pumped mid-infrared type-II phase-matching BaGa$_4$Se$_7$ optical parametric oscillator with a ZnGeP$_2$ amplifier. *Opt. Lett.* **2020**, *45*, 3805–3808. [CrossRef]
22. Brosnan, S.; Byer, R. Optical parametric oscillator threshold and linewidth studies. *IEEE J. Quantum Elect.* **1979**, *15*, 415–431. [CrossRef]
23. Guo, Y.; Li, Z.; Lei, Z.; Luo, X.; Yao, J.; Yang, C.; Wu, Y. Synthesis, Growth of crack-free large-size BaGa$_4$Se$_7$ crystal, and annealing studies. *Cryst. Growth Des.* **2019**, *19*, 1282–1287. [CrossRef]
24. Koechner, W. Transient thermal profile in optically pumped laser rods. *J. Appl. Phys.* **1973**, *44*, 3162–3170. [CrossRef]
25. He, Y.; Guo, Y.; Xu, D.; Wang, Y.; Zhu, X.; Yao, J.; Yan, C.; Tang, L.; Li, J.; Zhong, K.; et al. High energy and tunable mid-infrared source based on BaGa4Se7 crystal by single-pass difference-frequency generation. *Opt. Express* **2019**, *27*, 9241–9249. [CrossRef]
26. Tucker, J.; Marquardt, C.; Bowman, S.; Feldman, B. Transient thermal lens in a ZnGeP$_2$ crystal. *Appl. Optics* **1995**, *34*, 2678–2682. [CrossRef]
27. Yang, F.; Yao, J.; Xu, H.; Zhang, F.; Zhai, N.; Lin, Z.; Zong, N.; Peng, Q.; Zhang, J.; Cui, D.; et al. Midinfrared optical parametric amplifier with 6.4–11 μm range based on BaGa$_4$Se$_7$. *IEEE Photonics Technol. Lett.* **2015**, *27*, 1100–1103. [CrossRef]
28. Kolker, D.; Kostyukova, N.; Boyko, A.; Badikov, V.; Bsdikov, D.; Shadrintseva, A.; Tretyakova, N.; Zenov, K.; Karapuzikov, A.; Zondy, J. Widely tunable (2.6–10.4 μm) BaGa$_4$Se$_7$ optical parametric oscillator pumped by a Q-switched Nd:YLiF$_4$ laser. *J. Phys. Commun.* **2018**, *2*, 035039. [CrossRef]

 *applied sciences*

Article

A Reliable Way to Improve Electrochemical Migration (ECM) Resistance of Nanosilver Paste as a Bonding Material

Zikun Ding [1], Zhichao Wang [2], Bowen Zhang [3], Guo-Quan Lu [4] and Yun-Hui Mei [3,*]

1   School of Materials Science and Engineering, Tianjin University, Tianjin 300072, China; dingzikun970624@outlook.com
2   Beijing Century Goldray Semiconductor Co., Ltd., Beijing 100176, China; 18310709267@139.com
3   School of Electrical Engineering, Tiangong University, Tianjin 300350, China; bowenzhang@tiangong.edu.cn
4   The Bradley Department of Electrical and Computer Engineering, Virginia Tech, Blacksburg, VA 24061, USA; gqlu@vt.edu
*   Correspondence: meiyunhui@tiangong.edu.cn

Abstract: Electrochemical migration (ECM) of sintered nano-Ag could be a serious reliability concern for power devices with high-density packaging. An anti-ECM nano-Ag-SiO$_x$ paste was proposed by doping 0.1wt% SiO$_x$ nanoparticles rather than previously used expensive noble metals, e.g., palladium. The ECM lifetime of the sintered nano-Ag-SiO$_x$ was 1.5 to 3 times longer than that of the sintered nano-Ag, due to the fact that the SiO$_x$ could protect the Ag from oxidation. The thermo-mechanical reliability of the sintered nano-Ag-SiO$_x$ was also improved by sintering under 5 MPa assisted pressure. The lesser porosity and smaller grain boundaries of the sintered nano-Ag-SiO$_x$ could also be beneficial to retard the silver ECM. In the end, a double-sided semiconductor device was demonstrated to validate the better resistance to the ECM using the sintered nano-Ag-SiO$_x$.

Keywords: electrochemical migration; ionic migration; sintering; nanosilver; doping; Ag nanoparticles; reliability

Citation: Ding, Z.; Wang, Z.; Zhang, B.; Lu, G.-Q.; Mei, Y.-H. A Reliable Way to Improve Electrochemical Migration (ECM) Resistance of Nanosilver Paste as a Bonding Material. *Appl. Sci.* **2022**, *12*, 4748. https://doi.org/10.3390/app12094748

Academic Editor: David G. Calatayud

Received: 2 April 2022
Accepted: 6 May 2022
Published: 9 May 2022

Publisher's Note: MDPI stays neutral with regard to jurisdictional claims in published maps and institutional affiliations.

1. Introduction

Die-attach materials are crucial to power devices. It bonds the chip to the substrate and performs the functions of conducting current and heat. Traditional lead-free and Sn Pb solders harm the environment and have lower operating temperatures, thus nano-Ag paste as a die-attach material is a feasible option with excellent electrical and thermal conductivity [1–3].

However, silver is prone to electrochemical migration (ECM) and forms dendritic crystals, which result in short-circuit failure [4,5]. Generally, moisture is believed to play a crucial role in the migration of silver, and as a result, silver migration is described as a moisture migration phenomenon. However, recent studies have found that ECM failure of silver occurs even in high temperature and dry environments. The role of O$_2$ in dry environments is similar to that of H$_2$O in the classical moisture migration process [6]. Firstly, silver oxidized at the anode and form the intermediate species Ag$_2$O, which dissociated into silver cations and oxygen anions at a high temperature above 250 °C. Driven by the bias voltage, the silver cations move from anode to cathode through the insulating gap, while the oxygen anions move in the opposite direction to maintain charge neutrality. The continuous depletion of silver cations at the anode drives oxidation and dissociation until silver dendrites formed between the two electrodes. Subsequently, silver dendrites gradually appeared and result in short circuits [7,8].

It was concerned that the ECM of the sintered nano-Ag should be a failure risk for packaging wide band-gap (WBG) power semiconductors devices, which can theoretically be operated at extreme temperatures of 500 °C, not to mention the great increase of electrical bias [9]. Furthermore, the previous ECM of silver was only focused on thick film conductors in conventional printed circuit boards (PCBs) [4,10]. Double-sided cooling (DSC) silicon carbide (SiC) power devices have attracted more and more attention in industries, especially in electric vehicle applications, because of their superior heat dissipation capability and power density [11]. However, the DSC SiC devices have a very small electrode spacing, e.g., 250 μm [12]. When the great sintered nano-Ag was used as die attachment in the DSC SiC devices, it is crucial to avoid the ECM failure of sintered nano-Ag to guarantee the reliability of electric vehicles. Unfortunately, there is a current lack of research on the ECM of silver suppression for the DSC SiC power devices.

Inhibition of the ECM of silver, such as alloying silver with indium or copper to form anti-ECM Ag-In and Ag-Cu alloys, had been studied [13,14]. Recently, we proposed the method of alloying silver with palladium aimed at inhibiting the formation of silver ions through oxidation and decomposition in the anode to delay the ECM of silver [15]. However, these alloying methods need very high processing temperatures in order to improve the resistance to the ECM of silver by alloying more at the higher temperature, not to mention the extremely high cost of indium, palladium, and so on.

In this paper, a nano-Ag composite paste was proposed with the 0.1wt% SiO_x nanoparticles (NPs) as fillers. The conventional nano-Ag paste was compared as a reference. The lifetime for reaching the ECM failure was characterized in the first place. Then the thermomechanical reliability of the proposed nano-Ag-SiO_x paste was verified in the aspects of die-shearing strength and thermal impedance by thermal shocking tests because it is critical to ensure promising thermo-mechanical reliability besides the improvement of the resistance to the ECM. Microstructures and porosity were analyzed to clarify the effects of SiO_x doping on the ECM and the thermo-mechanical reliability. In the end, the proposed nano-Ag-SiO_x paste was used as a die attachment for a DSC power device. The improvement of the lifetime of the ECM was validated in the high power density demonstration.

2. Materials and Methods

2.1. Materials

The nano-Ag composite paste was made by ultrasonically mixing the 30 nm Ag NPs and the 30 nm SiO_x NPs with a specific weight ratio such as 0.1 wt%, accompanied by organic vehicles that included surfactants, binders, and thinners [16]. Figure 1a shows that the surfactant, binders, and thinners are dissolved by acetone in a beaker as a mixture. As seen from Figure 1b, in order to disperse the SiO_x NPs and the Ag NPs uniformly in the mixture, the mixture was stirred at 2000 rpm for 10 min using a planetary centrifugal mixer (THINKY ARE310).

2.2. Methods of Reliability Verification

Firstly, the capability of the sintered nano-Ag or nano-Ag-SiO_x against the ECM was verified in this work. The ECM samples were formed by stencil-printed the proposed anti-ECM nano-Ag composite paste on an alumina ceramic substrate (30×30 mm^2) to form pairs of electrodes with identical gaps of 0.5 mm for ensuring uniform electrical bias distributions (Figure 1c). The as-printed ECM samples were then heated to 280 °C at a ramp rate of 5 °C/min and sintered at 280 °C for 30 min. These processes are shown in Figure 1d. The leakage currents between the electrodes are monitored by a Pico ammeter (RIGOL DM3068). The time when the leakage current first reaches 1 mA was defined as the lifetime of the ECM. The environment temperature during the ECM test was 25 ± 2 °C and the environment humidity during the ECM test was 50%.

Then thermal shocking tests were carried out according to the standard of JESD22-A104C to verify the reliability of the nano-Ag composite paste. The temperature swung from −40 °C to +125 °C and the soaking time at the extreme temperatures was 10 min. Each period was about 25 min. The degradation of shear strength and thermal resistance of the ECM samples was recorded every 100 cycles until 1000 cycles. The shear strength was measured by a die-shearing tester (XYZTEC, CONDOR 150) and the thermal resistance was recorded by a thermal resistance measurement system that used V_{ge} as temperature sensitive parameter [17,18].

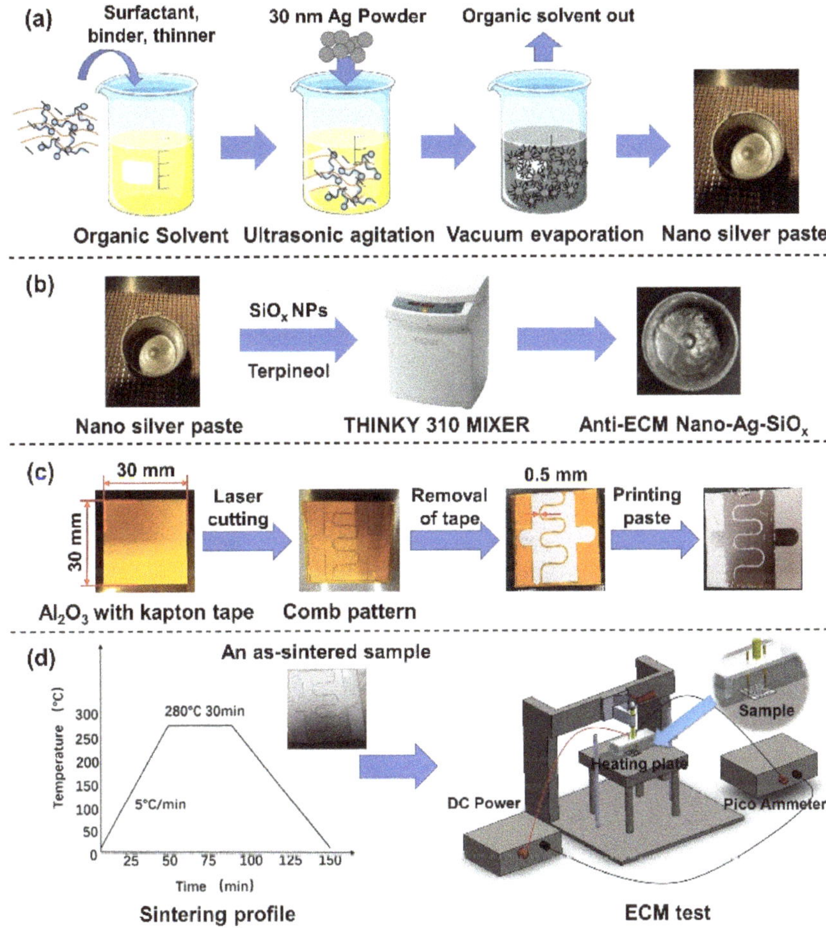

Figure 1. (**a**) Preparation process of nano silver paste; (**b**) preparation of proposed anti-ECM nano-Ag-SiO$_x$ composite paste; (**c**) preparation of ECM samples; (**d**) schematic of the sintering process and ECM test.

For the thermal resistance measurement, a $3.92 \times 3.88 \times 0.07$ mm^3 Insulated Gate-Bipolar-Transistor (IGBT) (INFINEON, IGC15T65QE) was sinter-bonded on a $30 \times 30 \times 1$ mm^3 silver-plated direct-bond-copper (DBC) substrate under the same process as that of the ECM samples.

For the die-shearing tests, a $3 \times 3 \times 0.25$ mm^3 dummy silicon chip with bottom metallization layer of Al-Ti-Ni-Ag (1.2 μm) was sinter-bonded on a $30 \times 30 \times 1$ mm^3 silver-plated DBC substrate using the same process as the ECM samples, under an additional pressure of 5 MPa.

In the end, a DSC power device was demonstrated to verify the anti-ECM of the proposed nano-Ag composite paste. The fabrication process of the DSC demonstration is shown in Figure 2. Firstly, the top side of a silicon carbide (SiC) dummy chip of $4 \times 4 \times 0.17$ mm^3 was metalized with Ti-Ni-Ag using magnetron sputtering technology. Then the proposed nano-Ag composite paste was printed onto two silver-plated DBC substrates of $20 \times 20 \times 1$ mm^3 and the top Ti-Ni-Ag metallization of the SiC chip separately. Then the chip, the silver-plated copper block buffer of $2 \times 2 \times 2.2$ mm^3, and the silver-plated power terminals were assembled with the DBC substrates. The as-assembled DBC substrates were sintered at 280 °C and 5 MPa as the process of the previous samples for the die-shearing tests.

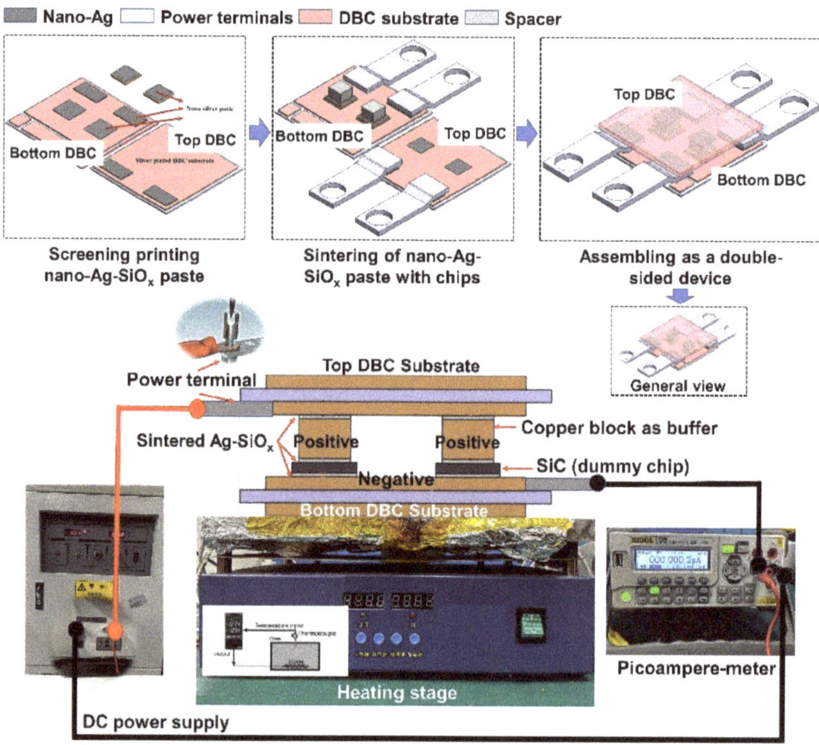

Figure 2. Fabrication process of the DSC demonstration and verification of electrochemical migration resistance for a DSC device using sintered nano-Ag-SiO$_x$ as die attachment.

During the ECM test for the demonstrated DSC power devices, a direct current (DC) power supplier (Topower TN-XX702) was connected with the power terminals of the device. A controllable heating plate was used to provide the ambient temperature required for the ECM test. The leakage currents in the device were monitored by the Pico ammeter (RIGOL DM3068).

3. Results and Discussion

The ECM experimental lifespan findings of the suggested nano-Ag composite paste and the control samples of conventional nano-Ag paste are listed in Table 1. According to Table 1, the nano-Ag-0.1%SiO$_x$ paste has a substantially longer failure lifetime than the nano-Ag paste under identical conditions. The 0.1wt% SiO$_x$ doping in the original nano-Ag paste serves a critical function in preventing the silver from migrating electrochemically. Furthermore, the lifetime reduces greatly when the temperature and the electric field are increased.

Table 1. ECM testing conditions and lifetime comparisons.

Temperature (K)	Electrode Spacing (mm)	Voltage (V)	Lifetime (min)	
			Sintered Nano-Ag	Sintered Nano-Ag-SiO$_x$
673	0.5	200	332	512
673	0.5	260	294	477
673	0.5	300	236	412
773	0.5	200	134	341
773	0.5	260	81	244
773	0.5	300	74	208

It was believed that the ECM behavior of the sintered nano-Ag at high temperatures initiates from the oxidation of the sintered anode and the decomposition of Ag$_2$O [6]. When the temperature is constant, the higher the electrode voltage applied, the faster the Ag$^+$ formed after anodic oxidation. These Ag ions would reduce at the negative electrode as metallic Ag under the action of the electric field, resulting in the accelerated growth of conductive silver dendrites from the cathode and extended to the anode. As a result, the leakage current can be detected between the anode and the cathode. When the voltage remains constant, the higher the ambient temperature, the faster the ions travel. Then the produced Ag$^+$ speeds up the creation of silver dendrites under the specific electric field. Therefore, the ECM lifetime of the sintered nano-Ag-SiO$_x$ in the temperature range of 573–673 K is 1.5 to 3 times that of the sintered nano-Ag. It was likely that the SiO$_x$ covered on the silver, as shown in Figure 3a,b, should be susceptible to oxidation. The oxidation electrode potential of SiO$_x$ oxidation to SiO$_2$, i.e., ~0.06 V_{SCE}, was reported as more than one time lower than that of Ag oxidation to Ag$_2$O, i.e., 0.13 V_{SCE} [7,19]. As a result, silver oxidation was reduced and the ECM was delayed.

Figure 3c shows the change in shear strength following temperature cycling. The shear strength falls as the number of cycles increases. It should be noted that the shear strength of the Ag-SiO$_x$ paste drops to less than 30 MPa after 600 cycles, but the shear strength of the nano-Ag control samples remains ~30 MPa after 1000 cycles. It was likely that the atomic diffusion rate of Ag into Ag, i.e., ~1.455 $\times$ 10^{-23} m^2/s [20], is much higher than that of Ag into SiO$_x$, i.e., ~8.76 $\times$ 10^{-29} m^2/s [21], resulting in an insufficient driving force for the sinter-bonding of Ag-SiO$_x$ heterogeneous particles under the same pressure-free and sintering temperature conditions [22]. The Young's modulus of the SiO$_x$ nanoparticles, on the other hand, is much higher than that of the Ag nanoparticles, and the thermal expansion coefficient is much lower than that of the sintered nano-Ag, so the interfacial bonds of the sintered Ag-SiO$_x$ were subjected to thermo-mechanical stress fatigue during temperature cycling, eventually leading to its shear strength dropping faster in the case of pressureless sintering [23].

It was known the driving force for the sintering can be increased greatly if assisted pressure was applied [24]. An extra 5 MPa pressure was applied to promote the sintering of the Ag-SiO$_x$ paste for better long-term reliability. Fortunately, the shear strength of the pressure-assisted sintered Ag-SiO$_x$ can be kept at ~35 MPa even after 1000 cycles. It was even higher than the shear strength of the nano-Ag control samples after 1000 cycles. It was believed that the pressure can ensure the better thermo-mechanical reliability of the nano-Ag-SiO$_x$ paste.

Figure 3c also shows the change in thermal resistance after the temperature cycling. The thermal resistance grows steadily as the number of cycles increases. The thermal resistance of the nano-Ag paste improves by ~9% after 1000 cycles, while the thermal resistance of the nano-Ag-SiO$_x$ paste sintered with 5 MPa pressure increases by ~12%. It was consistent with the variation of the die shear strength. Fortunately, the increase in thermal resistance of the nano-Ag-SiO$_x$ pastes is still less than the common industrial failure criterion in power electronics, which is usually required as less than 20% of the thermal resistance after 1000 cycles of the temperature cycling.

Figure 4 shows the microstructures and porosity of sintered Ag and sintered nano-Ag-SiO_x before and after the thermal shocking aging. The porosity of the sintered nano-Ag-SiO_x before and after the thermal shocking aging is always lower than that of the sintered nano-Ag. It was believed that the nucleation of metallic Ag occurs preferentially adjacent to the SiO_x nanoparticles as seeds; this is called heterogeneous nucleation [25,26]. Therefore, for example, the porosity of the as-sintered nano-Ag-SiO_x, i.e., 21.8%, can be much lower than that of the as-sintered nano-Ag, i.e., 34.5%.

Furthermore, the grains of the sintered nano-Ag-SiO_x grew more than that of the sintered nano-Ag under a similar sintering profile. The larger grain size indicated a smaller grain boundary density as well. Considering that the grain boundaries are close to the free surface as fast diffusion paths [27], the more the grain boundary, the more the Ag hillocks could be formed by atomic diffusion or stress migration through the grain boundaries. The fewer grain boundaries of the sintered nano-Ag-SiO_x may be also beneficial to retard the silver ECM to some extent as a result.

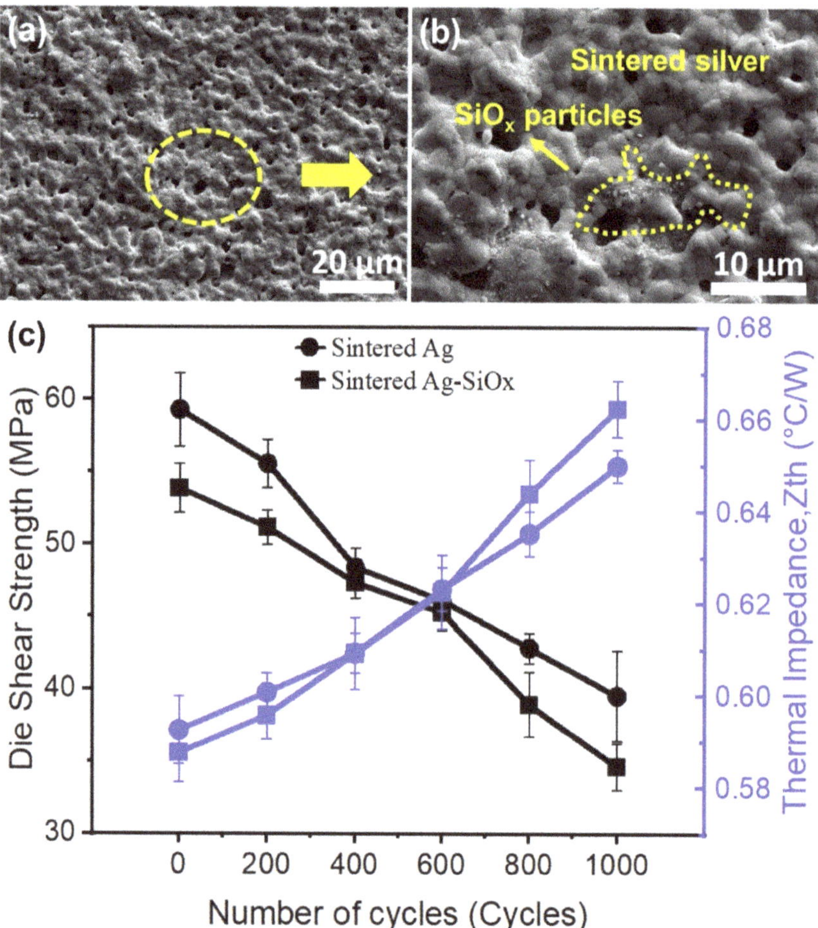

Figure 3. (**a**) Microstructures of as-sintered nano-Ag-SiO_x; (**b**) the corresponding magnified image; (**c**) the thermal impedance of sintered nano-Ag-SiO_x variation with the number of thermal shocking cycles.

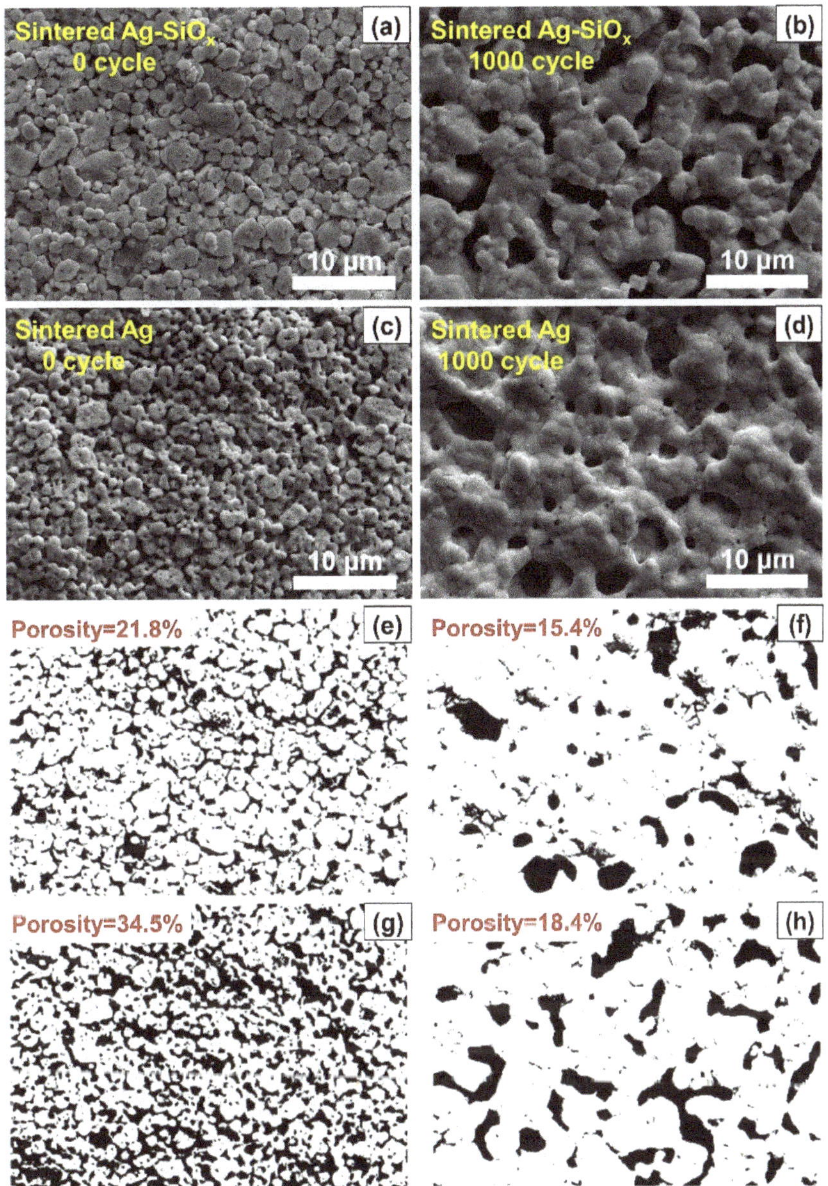

Figure 4. (**a–d**) Microstructures and (**e–h**) porosity comparisons of sintered nano-Ag and sintered nano-Ag-SiO$_x$ before and after the thermal shocking test.

For the ECM verification for device packaging applications at 400 °C under the applied voltage of 400 V, the ECM lifetime of the DSC device using the sintered nano-Ag-SiO$_x$ and the sintered nano-Ag as die attachment is 240 h and only 96 h, respectively. The ECM of silver was suppressed successfully. As seen from Figure 5, although silver dendrites appeared after 240 h ECM testing of DSC devices using sintered nano-Ag-SiO$_x$, significant improvement in ECM lifetime was achieved by nano-Ag-SiO$_x$ paste. Herein, the nano-Ag-SiO$_x$ paste could be a promising option to be used as a bonding material against specific metal corrosion.

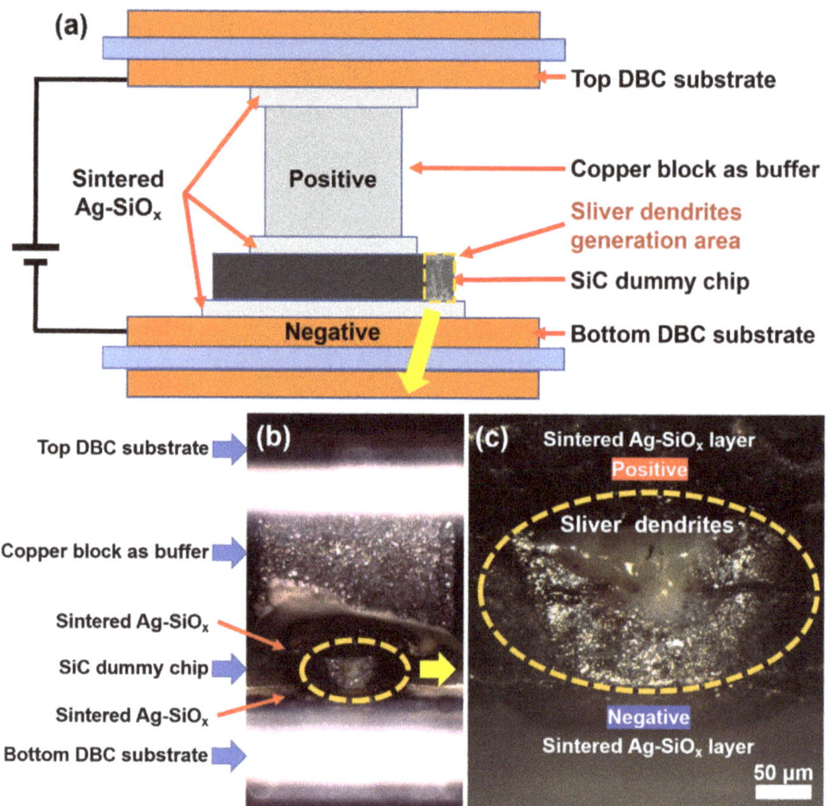

Figure 5. (**a**) Schematic diagram of DSC device, (**b**) metallographic micrograph of the DSC device after ECM failure, and (**c**) the corresponding magnified image.

4. Conclusions

Silver ECM at high temperatures could be significantly retarded by the addition of SiO_x nanoparticles in the nano-Ag paste. The thermo-mechanical reliability of the proposed nano-Ag-SiO_x paste has been verified and proved to be improved by sintering under 5 MPa assisted pressures. It seems suitable as a new promising bonding material to prevent silver ECM failure because its die shear strength and thermal resistance degradation still did not reach the common failure criterion even after 1000 cycles of the thermal shock test. Finally, the ECM verification of the DSC power device further confirms the improvement of the sintered nano-Ag-SiO_x as die attachment against the sintered nano-Ag ECM failure at high temperatures. It is interesting to pay more attention to reducing the ECM failure of sintered nano-Ag in economical and promising ways in the future.

Author Contributions: Conceptualization, Y.-H.M. and G.-Q.L.; methodology, Y.-H.M. and B.Z.; validation, Z.D.; formal analysis, Z.D., Z.W. and Y.-H.M.; investigation, Z.D. and Y.-H.M.; resources, Y.-H.M. and Z.W.; data curation, B.Z. and Y.-H.M.; writing—original draft preparation, Z.D. and Y.-H.M.; writing—review and editing, Y.-H.M. and B.Z.; visualization, Z.D. and Y.-H.M.; supervision, Y.-H.M.; project administration, Y.-H.M.; funding acquisition, Y.-H.M. and Z.W. All authors have read and agreed to the published version of the manuscript.

Funding: This research was funded by the National Natural Science Foundation of China, grant numbers 52177189, 51922075, and 52107203, and by Tianjin Municipal Science and Technology Bureau, grant number 21JCJQJC00150. The APC was funded by the National Natural Science Foundation of China.

Institutional Review Board Statement: Not applicable.

Informed Consent Statement: Not applicable.

Data Availability Statement: Not applicable.

Acknowledgments: In this section, you can acknowledge the equipment support given by Xin Li and her students from Tianjin University.

Conflicts of Interest: The authors declare no conflict of interest.

References

1. Wang, T.; Chen, X.; Lu, G.Q.; Lei, G.Y. Low-Temperature Sintering with Nano-Silver Paste in Die-Attached Interconnection. *J. Electron. Mater.* **2007**, *36*, 1333–1340. [CrossRef]
2. Matkowski, P.K.; Falat, T.; Zaluk, Z.; Felba, J.; Moscicki, A. Structure of the Thermal Interface Connection Made of Sintered Nano Silver. In Proceedings of the 2014 IEEE Electronics System-Integration Technology Conference, Dresden, Germany, 7–11 May 2014; pp. 1–6.
3. Yang, C.A.; Kao, C.R.; Nishikawa, H. Development of Die Attachment Technology for Power IC Module by Introducing Indium into Sintered Nano-Silver Joint. In Proceedings of the 2017 IEEE 67th Electronic Components and Technology Conference (ECTC), Orlando, FL, USA, 30 May–2 June 2017; pp. 2377–5726.
4. Kim, K.S.; Bang, J.O.; Jung, S.B. Electrochemical Migration Behavior of Silver Nanopaste Screen-Printed for Flexible and Printable Electronics. *Curr. Appl. Phys.* **2013**, *13*, S190–S194. [CrossRef]
5. Yang, S.; Christou, A. Failure Model for Silver Electrochemical Migration. *IEEE Trans. Device Mater. Reliab.* **2007**, *7*, 188–196. [CrossRef]
6. Noh, B.I.; Yoon, J.W.; Kim, K.S.; Lee, Y.C.; Jung, S.B. Microstructure, Electrical Properties, and Electrochemical Migration of a Directly Printed Ag Pattern. *J. Electron. Mater.* **2011**, *40*, 35–41. [CrossRef]
7. Li, D.; Mei, Y.H.; Tian, Y.; Xin, Y. Doping Low-Cost SiO_x (1.2 < x < 1.6) Nanoparticles to Inhibit Electrochemical Migration of Sintered Silver at High Temperatures. *J. Alloys Compd.* **2020**, *830*, 154587.
8. Medgyes, B.; Illés, B.; Harsányi, G. Electrochemical Migration Behaviour of Cu, Sn, Ag and Sn63/Pb37. *J. Mater. Sci. Mater. Electron.* **2012**, *23*, 551–556. [CrossRef]
9. Mei, Y.H.; Lu, G.Q.; Chen, X.; Luo, S.; Ibitayo, D. Migration of Sintered Nanosilver Die-Attach Material on Alumina Substrate Between 250 °C and 400 °C in Dry Air. *IEEE Trans. Device Mater. Reliab.* **2011**, *11*, 316–322. [CrossRef]
10. Mcnutt, T.; Passmore, B.; Fraley, J.; Mcpherson, B.; Shaw, R.; Olejniczak, K.; Lostetter, A. High-Performance, Wide-Bandgap Power Electronics. *J. Electron. Mater.* **2014**, *43*, 4552–4559. [CrossRef]
11. Zhou, Y.; Huo, Y. The Comparison of Electrochemical Migration Mechanism between Electroless Silver Plating and Silver Electroplating. *J. Mater. Sci. Mater. Electron.* **2016**, *27*, 931–941. [CrossRef]
12. Liu, X.; Wu, Z.; Yan, Y.; Kang, Y.; Chen, C. A Novel Double-Sided Cooling Inverter Leg for High Power Density EV Based on Customized SiC Power Module. In Proceedings of the 2020 IEEE Energy Conversion Congress and Exposition (ECCE), Detroit, MI, USA, 11–15 October 2020; pp. 3151–3154.
13. Cheng, T.H.; Nishiguchi, K.; Fukawa, Y.; Baliga, B.J.; Hopkins, D.C. Characterization of Highly Thermally Conductive Organic Substrates for a Double-Sided Cooled Power Module. *Int. Symp. Microelectron.* **2020**, *1*, 277–281. [CrossRef]
14. Yang, C.A.; Wu, J.; Lee, C.C.; Kao, C.R. Analyses and Design for Electrochemical Migration Suppression by Alloying Indium into Silver. *J. Mater. Sci. Mater. Electron.* **2018**, *29*, 13878–13888. [CrossRef]
15. Zhang, D.; Zou, G.; Liu, L.; Zhang, Y.; Yu, C.; Bai, H.; Zhou, N. Synthesis with Glucose Reduction Method and Low Temperature Sintering of Ag-Cu Alloy Nanoparticle Pastes for Electronic Packaging. *Jpn. Inst. Met. Mater.* **2015**, *56*, 1252–1256. [CrossRef]
16. Wang, D.; Mei, Y.H.; Xie, H.; Zhang, K.; Siow, K.S.; Li, X.; Lu, G.Q. Roles of Palladium Particles in Enhancing the Electrochemical Migration Resistance of Sintered Nano-Silver Paste as a Bonding Material. *Mater. Lett.* **2017**, *206*, 1–4. [CrossRef]
17. Bai, J.G.; Calata, J.N.; Lu, G.Q. Processing and Characterization of Nanosilver Pastes for Die-Attaching SiC Devices. *IEEE Trans. Electron. Packag. Manuf.* **2007**, *30*, 241–245. [CrossRef]
18. Gang, C.; Dan, H.; Mei, Y.H.; Cao, X.; Tao, W.; Xu, C.; Lu, G.Q. Transient Thermal Performance of IGBT Power Modules Attached by Low-Temperature Sintered Nanosilver. *IEEE Trans. Device Mater. Reliab.* **2012**, *12*, 124–132.
19. Mei, Y.; Tao, W.; Xiao, C.; Gang, C.; Lu, G.Q.; Xu, C. Transient Thermal Impedance Measurements on Low-Temperature-Sintered Nanoscale Silver Joints. *J. Electron. Mater.* **2012**, *41*, 3152–3160. [CrossRef]
20. Li, D.; Mei, Y.; Xin, Y.; Li, Z.; Chu, P.K.; Ma, C.; Lu, G.Q. Reducing Migration of Sintered Ag for Power Devices Operating at High Temperature. *IEEE Trans. Power Electron.* **2020**, *35*, 12646–12650. [CrossRef]
21. Wazzan, A.R.; Tung, p.; Robinson, L.B. Diffusion of Silver into Nickel Single Crystals. *J. Appl. Phys.* **1971**, *42*, 5316–5320. [CrossRef]
22. Friedland, E.; Malherbe, J.B.; van der Berg, N.G.; Hlatshwayo, T.; Botha, A.J.; Wendler, E.; Wesch, W. Study of Silver Diffusion in Silicon Carbide. *J. Nucl. Mater.* **2009**, *389*, 326–331. [CrossRef]
23. Hu, B.; Yang, F.; Peng, Y.; Hang, C.; Chen, H.; Lee, C.; Yang, S.; Li, M. Effect of SiC Reinforcement on the Reliability of Ag Nanoparticle Paste for High-Temperature Applications. *J. Mater. Sci. Mater. Electron.* **2019**, *30*, 2413–2418. [CrossRef]

24. Zhang, H.; Chen, C.; Nagao, S.; Suganuma, K. Thermal Fatigue Behavior of Silicon-Carbide-Doped Silver Microflake Sinter Joints for Die Attachment in Silicon/Silicon Carbide Power Devices. *J. Electron. Mater.* **2017**, *46*, 1055–1060. [CrossRef]
25. Yan, H.; Liang, P.; Mei, Y.H.; Feng, Z. Brief Review of Silver Sinter-bonding Processing for Packaging High-temperature Power Devices. *Chin. J. Electr. Eng.* **2020**, *6*, 25–34. [CrossRef]
26. Li, W.; Hu, D.; Li, L.; Li, C.F.; Jiu, J.; Chen, C.; Ishina, T.; Sugahara, T.; Suganuma, K. Printable and Flexible Copper-Silver Alloy Electrodes with High Conductivity and Ultrahigh Oxidation Resistance. *ACS Appl. Mater. Inter.* **2017**, *9*, 24711–24721. [CrossRef] [PubMed]
27. Lin, S.K.; Nagao, S.; Yokoi, E.; Oh, C.; Zhang, H.; Liu, Y.C.; Lin, S.G.; Suganuma, K. Nano-Volcanic Eruption of Silver. *Sci. Rep.* **2016**, *6*, 34769. [CrossRef]

Article

Sparse Weighting for Pyramid Pooling-Based SAR Image Target Recognition

Shaona Wang *, Yang Liu and Linlin Li

Tianjin Key Laboratory of Optoelectronic Detection Technology and Systems,
School of Electrical and Electronic Engineering, Tiangong University, Tianjin 300387, China;
2030070862@tiangong.edu.cn (Y.L.); 2031070932@tiangong.edu.cn (L.L.)
* Correspondence: wangshaona@tiangong.edu.cn

Abstract: In this study, a novel feature learning method for synthetic aperture radar (SAR) image automatic target recognition is presented. It is based on spatial pyramid matching (SPM), which represents an image by concatenating the pooling feature vectors that are obtained from different resolution sub-regions. This method exploits the dependability of obtaining the weighted pooling features generated from SPM sub-regions. The dependability is determined by the residuals obtained from sparse representation. This method aims at enhancing the weights of the pooling features generated in the sub-regions located in the target and suppressing the weights of the background. The feature representation for SAR image target recognition is discriminative and robust to speckle noise and background clutter. Experiments performed on the Moving and Stationary Target Acquisition and Recognition public dataset prove the advantageous performance of the presented algorithm over several state-of-the-art methods.

Keywords: synthetic aperture radar (SAR) images; automatic target recognition (ATR); spatial pyramid matching (SPM); sparse representation; pooling

Citation: Wang, S.; Liu, Y.; Li, L. Sparse Weighting for Pyramid Pooling-Based SAR Image Target Recognition. *Appl. Sci.* **2022**, *12*, 3588. https://doi.org/10.3390/app12073588

Academic Editors: Pingjuan Niu, Li Pei, Yunhui Mei, Hua Bai and Jia Shi

Received: 4 March 2022
Accepted: 29 March 2022
Published: 1 April 2022

Publisher's Note: MDPI stays neutral with regard to jurisdictional claims in published maps and institutional affiliations.

1. Introduction

Synthetic aperture radar (SAR) imagery has become a significant research object in many areas, such as civilian and military fields [1–3]. It takes advantage of acquiring images in all weather conditions and during the night as well as the day. Image target recognition is a basic step in understanding and interpreting SAR images [4]. In this context, it is important to develop discriminative and robust methods for automatic target recognition (ATR) systems, and tremendous research attention has been paid to the study of ATR for SAR images [5–9].

Recently, sparse representation (SR) has become a focus and has been used in many areas [10]. It is robust to noise and can maintain natural discrimination without any prior information. Even in SAR target recognition, SR could remove the need for the pose estimating process. Recently, Thiagarajan et al. [11] applied SR to realize SAR image target recognition and applied a local linear approximation to generate a classification prediction for every target class manifold. This algorithm did not demand any specific pose estimation or preprocessing, but the use of random projections in the high-dimensional space discarded some discriminative locality information, thus making occlusion handling more difficult. In another study [12], descriptors from local patches were extracted, and an image was treated as a collection of the unordered descriptors; then, sparse representation was applied to represent the local patches for SAR image target classification in the framework of SPM. Additionally, the application of spatial pyramids was confirmed to be an effective classification method for SAR images.

This model of spatial pyramid matching [13] for image classification is a statistics-based model with the objective of providing a better image representation. In order to obtain discriminant details of the images, the SPM needs to extract the low-level local features through, for example, scale-invariant feature transform (SIFT) [14] and histogram of oriented gradient (HOG) [15]. However, the local features are not provided directly to image classifiers due to their sensitivity to noise and computational complexity. One solution is to represent the images by integrating the local features into the midlevel features. This image representation works well with linear classifiers, and the results have achieved a competitive performance in many image classification tasks. Nonetheless, the SPM model is not perfect when implemented on SAR images. This is attributed to the fact that the variety of targets posed in SAR images hinders the advantages of SPM. Noteworthy, locality is more essential than sparsity [16], locality-constrained linear coding was presented in place of the vector quantization (VQ) coding, and a good approximation was obtained. Recently, Zhang et al. [17] proposed to apply a locality constraint to ensure similar patches shared similar codes in the coding scheme for SAR image target recognition. However, its complete codebook was obtained after preprocessing of the estimation of target poses.

In a literature study [5], a complementary spatial pyramid coding method was used in change detection, and good performance was gained. The SAR targets suffer from the effects of the speckle noise and the background; therefore, different parts of an image play different roles in image representation. Combining the advantages of SPM and SR, a novel SAR image target recognition method is proposed herein, which makes use of the dependability obtained by SR to obtain weighted sub-regions at every pyramid level. Some sub-regions in each level of the spatial pyramid may consist of background noise, and other ones represent the target. Based on the sparse representation theory, the target parts could be represented by the training samples from the same class [18]. Therefore, there could be a small residual value corresponding to its class, which shows the dependability of the sub-region. We apply the dependability to weight the pooling features obtained from the SPM sub-regions [19]. Therefore, the pooling feature located in the target is enhanced. Meanwhile, the pooling feature located in the background is suppressed. The results obtained using real SAR Moving and Stationary Target Acquisition and Recognition (MSTAR) database demonstrate that the method presented herein is more robust to variant unconstrained conditions than the methods reported in other recent related studies.

The organization of this paper is as follows: Section 2 introduces the presented sub-regions weighting method. Section 3 reports the experimental results of the presented algorithm and compares the approach used herein with some classical approaches. Finally, Section 4 includes the conclusion.

2. SAR Image Recognition with Sparse Weighting Spatial Pyramid Pooling

The method proposed herein simultaneously utilizes the SPM model and SR to deal with SAR image recognition. Enlightened by the idea that different parts of an image play different roles, a sparse weighting spatial pyramid pooling method is proposed to extract a new type of feature. The main objective of utilizing this method is to reduce the influence of background clutter and enhance the target. Figure 1 exhibits the flowchart of the proposed SAR image target recognition method. Firstly, an image was divided into gradual fine sub-regions; then, the dense local features were calculated, followed by coding and pooling according to SPM, in order to obtain feature vectors of the sub-regions, respectively. Additionally, the pooling feature vector of the sub-regions at each pyramid level was weighted based on the dependability, which was determined according to the residuals obtained by the SR. Finally, the representation for SAR images was built by systematically concatenating the weighted feature vectors. With sparse representation classification, the method is robust, in particular, in dealing with the speckle noise and large clutter background.

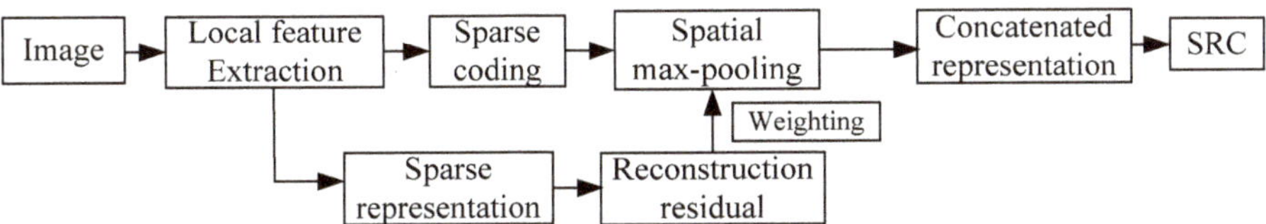

Figure 1. Overview of the SAR image target recognition flowchart.

2.1. Local Feature Extraction and Descriptor Quantization

Considering that the SIFT feature has the characteristic of invariance to scale, orientation, and affine distortion, in this study, dense-SIFT descriptors were extracted to express an SAR image [14]. The image I can be denoted by a set of local features' descriptors as: $I = [d_1, d_2, \ldots, d_N] \in \mathrm{R}^{D \times N}$, where d_i represents the $D-$dimensional SIFT description vector. Given a codebook $B = [b_1, b_2, \ldots, b_M] \in \mathrm{R}^{D \times N}$, the K-Means clustering algorithm [13] was adopted with the Euclidean distance to cluster local features into groups, and the generated centers of each group were taken as the codebook.

Sparse coding [10] was applied to encode the local feature vectors, as follows:

$$\arg \min_V \sum_{i=1}^{N} ||d_i - Bv_i||_2 + \lambda ||v_i||_1. \tag{1}$$

The feature vector for a descriptor d_i becomes an M-dimensional vector v_i. v_i is the corresponding vector to the descriptor d_i. The sparse coding vectors are obtained by the feature-sign search algorithm [20] by solving Equation (1) here.

The next step is the spatial pooling step aiming at obtaining more discriminative image representation from each sub-region. This study applied the construction of a three-level spatial pyramid. At every resolution e, $e = 0, 1, 2$, a grid was constructed such that there were 2^e resolution cells along every dimension, i.e., the $1 \times 1, 2 \times 2, 4 \times 4$ grid structures; thus, the sub-regions were $K = 21$ in total. We let V be a collection containing T local feature codes acquired from a sub-region. In addition, the max-pooling strategy was used to concatenate all the descriptors for the sub-region, which can acquire many discriminative features to SAR image spatial variations. The expression of the max-pooling is as follows:

$$t_k = \max(v_1, v_2, \cdots, v_T) \tag{2}$$

where max represents the element-wisely maximization for the involved vectors. The local features were pooled in all the sub-regions and concatenated to form the image representation $f = [t_1; t_2; \cdots; t_K]$.

Considering a set of G training images from C classes, $F = [f_1, f_2, \cdots, f_G]$, where f_g is the feature vector, which is the pooling result of the gth image. Correspondingly, F is partitioned as $F = \left[F_1^T, F_2^T, \cdots, F_K^T \right]^T$ and $F_k \in R^{d \times G}$ with $d < G$. Distinctly, the gth column of F_k is the feature vector for the kth sub-region of the gth image, and G represents the image number of the training dataset. Analogously, a test image y is divided into $y = [y_1; y_2; \cdots; y_K]$.

2.2. Determination of Sub-Region Weights

The traditional SPM models use all the extracted local descriptors for image representation; therefore, some of them might not be the objects to be recognized. Thus, it was required to learn the spatial weights to boost the target part and weaken the background parts. Taking the sparse representation into account, the observation to be followed was that the target part was sparser than the background one. The feature vectors of the sub-regions at every pyramid level were weighted on the residual basis to determine their dependability [21]. Then, the following optimization problem [10] was obtained:

$$\hat{x}_k = \text{argmin}_{x_k} \left\{ ||F_k x_k - y_k||_2^2 + \lambda ||x_k||_1 \right\} \text{ s.t. } 1 \leq k \leq K ,\qquad(3)$$

where y_k is the pooling feature for each sub-region, and F_k is the dictionary.

The residuals of a sub-region can be measured simply by using the L2-norm as follows:

$$r_{kc} = ||F_k \delta_c(\hat{x}_k) - y_k||_2,\qquad(4)$$

where r_{kc} is the residual corresponding to the cth class of the kth sub-region, $\delta_c(\hat{x}_k)$ is the discriminative function, which selects the elements related with the cth class in $\hat{x}_k$, $c = [1, 2, \cdots, C]$, and C is the class number of the training dataset.

All the residuals of the kth sub-region were concatenated to generate the residual vector r_k as follows:

$$r_k = [r_{k1}, r_{k2}, \cdots, r_{kC}].\qquad(5)$$

According to the theory of SR [10], target sub-regions can be accurately represented only by the training samples from the same class. Therefore, the residual vector produced contains only one small element. The background sub-regions could be far away from every subspace spanned by the training samples of each class but nearly within the subspace spanned by all training samples of all classes. The residual vector produced could include almost similar elements. According to the abovementioned analysis, the following function was applied for evaluating the sub-region sparsity:

$$\varsigma_k = \frac{\min(r_k)}{\text{mean}(r_k)}.\qquad(6)$$

When r_k has only a single zero or near zero residual, ς_k reaches the minimum value 0. This shows that the sub-region could be well represented by its subspace. When all residuals in r_k are nonzero value and equal, ς_k reaches the maximum value 1. This indicates that the sub-region could contain noise or it could be the background part.

This is verified by the numerical value results shown in Figure 2. Figure 2a,b are example images, which illustrate the distribution of sub-regions in the construction of a three-level pyramid. Figure 2c,d show ς_k of the sub-regions corresponding to Figure 2a,b respectively. The obtained residuals indicated that the residuals' sparsity ς_k generated in the sub-regions located in the target parts (such as the sub-regions marked by 11 and 12 in Figure 2a) was smaller than the background parts (such as the sub-regions marked by 9 and 10 in Figure 2a) in the third-level pyramid. Additionally, a difference existed between the residual values in the different resolutions, such that the residuals were much greater in the low resolution level than in the high ones, in particular, the target parts. The reason could be that more background cluster and speckle noise was included in a larger region. Therefore, in order to differentiate the target sub-regions from the background ones, the residual could be employed to obtain the weighted feature vector.

To tackle this problem, a simple way was to impair the feature representation of the background sub-regions for the classification. This was achieved by weighting all sub-regions with the reciprocal of the sparsity: ω_k, which is viewed as the dependability:

$$\omega_k = \frac{1}{\varsigma_k}\qquad(7)$$

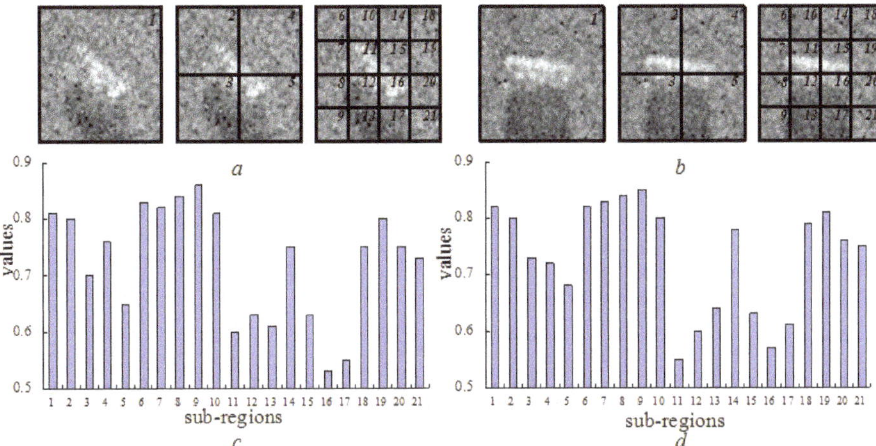

Figure 2. Illustration of the numeric value results of the sub-region sparsity. (**a**,**b**) are example images illustrating the distribution of sub-regions in the construction of the three-level pyramid. (**c**,**d**) show the ς_k of the sub-regions corresponding to the images (**a**,**b**).

2.3. Weighted SPM Sparse Representation

After weighting the sub-regions, the weighted sub-regions were integrated to reconstruct a global feature vector according to the SPM model. Each sub-region vector y_k was weighted by a corresponding weight to be represented as a sub-region weighted query vector:

$$y_k^{\omega} = \omega_k y_k. \tag{8}$$

We connected all sub-regions' weighted query vectors in series to produce a weighted feature vector as

$$y^{\omega} = \left[y_1^{\omega T}, y_2^{\omega T}, \cdots, y_K^{\omega T} \right]^T. \tag{9}$$

Consider that y^{ω} can be represented by F in a linear manner as follows:

$$y^{\omega} = Fu^{\omega}. \tag{10}$$

To obtain the primary label information from the training dataset matrix and weighted query feature vector, the SR classification method [13] was applied to complete the target image classification:

$$\hat{u}^{\omega} = \mathrm{argmin}_{u^{\omega}} \left\{ ||Fu^{\omega} - y^{\omega}||_2^2 + \lambda ||u^{\omega}||_1 \right\}. \tag{11}$$

Linear approximation [15] is capable of selecting significant training vectors to provide a good discrimination among all the classes.

2.4. Classification Procedure

After the representation coefficient $\hat{u}^{\omega}$ was obtained, the given weighted global query vector y^{ω} and the global training matrix F were applied to calculate the minimum reconstruction residual; then, the query image to the subject c that had the global minimum residual [12] was allocated:

$$c = \mathrm{argmin}_j ||F\delta_j(\hat{u}^{\omega}) - y^{\omega}||_2. \tag{12}$$

3. Contrast Experiment Results and Analysis

3.1. Dataset and Experimental Conditions

This section presents the experimental results on the MSTAR [22] public database to assess the performance of the presented algorithm. Three types of vehicles are included in the dataset. They are BMP2 armored personnel carriers and BTR70 and T72 main battle tanks with depression angles at 17 degrees and 15 degrees. Every image in the dataset is 128×128 pixels, covering the full 0–to–360 degree aspect range. The resolution of each image is 0.3 m $\times$ 0.3 m. The training set contained the SAR target images whose depression was 17 degrees. The other images with a depression of 15 degrees were regarded as the testing set. The visible light images are illustrated in Figure 3. Table 1 lists the details of the training set and testing set.

(a)

(b)

(c)

Figure 3. Visible light images for three targets in the MSTAR database. (**a**) BMP2. (**b**) BTR70. (**c**) T72.

Table 1. The dataset of training and testing targets.

Target	Name of Training Set	Training Set	Name of Testing Set	Testing Set
BTR70	Sn-c71	233	Sn-c71	196
BMP2	Sn-9663	233	Sn-9663	195
			Sn-9666	196
			Sn-c21	196
T72	Sn-132	232	Sn-132	196
			Sn-812	195
			Sn-s7	191

To reduce the negative interaction of the redundant background, a 64×64 pixels sub-image was cropped from the center of every image. Moreover, the amplitude of the sub-image was normalized. Herein, the experiment result, which is reported, was the average classification result of running 10 times.

In this experiment, the SIFT [14] was applied as the local feature. We extracted the SIFT features from patches densely located by six pixels on the image, under the scales of 16×16. Furthermore, every SIFT descriptor had the dimension of 128.

3.2. Experiment 1: Investigation of the Proposed Method

In order to verify the superiority of the presented algorithm, the presented sparse weighting sparse pyramid pooling (SWSPP) algorithm was compared with the state-of-the-art methods: namely sparse coding SPM (ScSPM) [13] and locality-constrained linear coding SPM (LLC) [16]. The classification results are listed in Table 2. The performance of the codebook was compared under different dimensions *dim* from 100 to 1024. The greater the dimensionality, the more the feature extracted was discriminative. Therefore, the results improved as the dimension increased. In all three scenarios, the method presented in this study was obviously superior to the ScSPM and LLC methods under all the dimensions of the codebook. Since the proposed SWSPP took the sub-regions dependability into account it enhanced the discrimination of the recognition target. From the experiment results, we verified that the dependability generated according to the sparse representation was effective for classification. The curves of the classification average results under the three algorithms are shown in Figure 4.

Table 2. MSTAR classification results (%).

dim	Data	ScSPM	LLC-SPM	SWSPP
100	BMP2	77.27	76.17	80.99
	BTR70	92.35	95.77	97.04
	T72	85.36	88.24	85.02
	av	84.99	86.72	87.68
200	BMP2	79.66	80.82	83.79
	BTR70	94.13	95.36	98.03
	T72	86.17	88.09	87.29
	av	86.65	88.09	89.70
400	BMP2	83.65	83.48	87.67
	BTR70	96.73	96.73	99.39
	T72	86.87	87.79	88.14
	av	89.08	89.33	91.73
600	BMP2	84.16	85.33	89.27
	BTR70	97.19	97.86	100
	T72	87.63	88.36	89.00
	av	89.66	90.52	92.76
800	BMP2	84.67	86.51	91.31
	BTR70	98.16	97.81	100
	T72	88.38	88.95	90.03
	av	90.41	91.09	93.78
1024	BMP2	87.05	86.46	90.46
	BTR70	96.94	98.78	100
	T72	89.18	89.78	92.27
	av	91.06	91.67	94.24

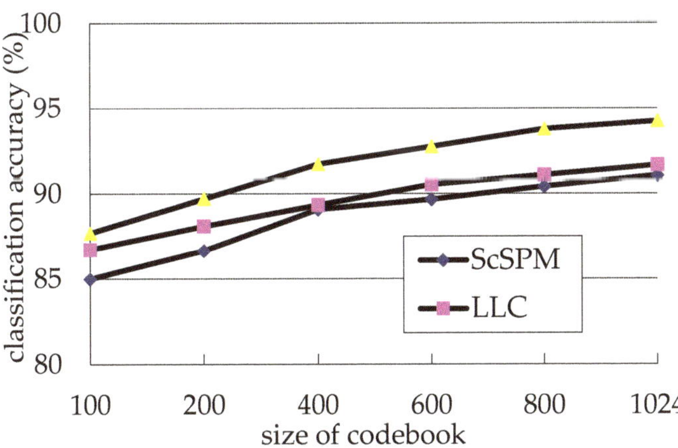

Figure 4. Curves of classification results of different codebook dimensions of the three different methods.

3.3. Experiment 2: Comparison of SWAPP with Related Algorithms

In order to show a more rounded analysis, the presented approach was compared with the following methods: the sparse representation classification (SRC) method [11], the PCA feature [23], and the LDA feature [24]. Table 3 gives the average classification results.

According to this table, we can see that the presented method had advantages over all the other methods. The average classification result of our proposed method was 2% higher than the other approaches. On the basis of the results, we can see that the classification accuracy of the presented method was better; in particular, the correct rate of the BTR70 was 100%. The result validates the effectiveness of the proposed dependability weighted feature learning method for SAR image recognition.

Table 3. Classification results (%) of different methods.

Algorithm	BMP2	BTR70	T72	Average
SRC [11]	84.48	98.33	94.59	92.47
PCA [23]	94.54	85.66	96.79	92.33
LDA [24]	98.47	67.80	95.86	87.38
SWSPP	90.46	100	92.27	94.24

4. Conclusions

In this study, a robust synthetic aperture radar (SAR) automatic target recognition approach was presented that applied sparse representation to produce a weighted global feature vector based on spatial pyramid matching. SAR image target recognition was performed on the reconstructed image representation. The experimental results clearly and consistently showed that the proposed framework was significantly more discriminative than the traditional SPM methods. Moreover, the feature coding–pooling framework was shown to perform well in image classification tasks, the unavoidable information loss incurred by the feature quantization in the coding process and the undesired dependence of pooling on the image spatial layout, however, may severely limit the recognition performance. Therefore, more systematic investigations are still required to find a new coding method to improve the classification performance in future research.

Author Contributions: Conceptualization, S.W. and Y.L.; methodology, S.W.; software, Y.L.; validation, L.L.; formal analysis, S.W.; investigation, L.L.; resources, S.W.; data curation, S.W.; writing—original draft preparation, Y.L.; writing—review and editing, S.W.; visualization, L.L.; supervision, S.W.; project administration, S.W.; funding acquisition, S.W. All authors have read and agreed to the published version of the manuscript.

Funding: This research was funded in part by the National Natural Science Foundation of China (Grant No. 61901297) and in part by the National Science Foundation of Tianjin Province of China, (Grant No. 18JCQNJC70600).

Institutional Review Board Statement: Not applicable.

Informed Consent Statement: Not applicable.

Data Availability Statement: Not applicable.

Conflicts of Interest: The authors declare no conflict of interest.

References

1. Luo, Z.; Jiang, X.; Liu, X. Synthetic minority class data by generative adversarial network for imbalanced sar target recognition. In Proceedings of the IGARSS 2020—2020 IEEE International Geoscience and Remote Sensing Symposium, Waikoloa, HI, USA, 26 September–2 October 2020; pp. 2459–2462.
2. Martone, A.; Innocenti, R.; Ranney, K. *Moving Target Indication for Transparent Urban Structures*; U.S. Army Research Laboratory: Adelphi, MD, USA, 2009.
3. Zhang, J.; Xing, M.; Xie, Y. FEC: A Feature Fusion Framework for SAR Target Recognition Based on Electromagnetic Scattering Features and Deep CNN Features. *IEEE Trans. Geosci. Remote Sens.* **2021**, *59*, 2174–2187. [CrossRef]
4. Mishra, A.K.; Bernard, M. Automatic target recognition using multipolar bistatic synthetic aperture radar images. *IEEE Trans. Aerosp. Electron. Syst.* **2010**, *46*, 1906–1920. [CrossRef]
5. Wang, S.; Jiao, L.; Yang, S.; Liu, H. SAR image target recognition via complementary spatial pyramid coding. *Neurocomputing* **2016**, *196*, 125–132. [CrossRef]

6.   O'Sullivan, J.A.; DeVore, M.D.; Kedia, V. SAR ATR performance using a conditionally Gaussian model. *IEEE Trans. Aerosp. Electron. Syst.* **2001**, *37*, 91–108. [CrossRef]

7.   Owirka, G.J.; Verbout, S.M.; Novak, L.M. Template-based SAR ATR performance using different image enhancement techniques. *Proc. SPIE* **1999**, *3721*, 302–319.

8.   Chen, Y.; Blasch, E.; Qian, T. Experimental feature-based SAR ATR perormance evaluation under different operational conditions. In Proceedings of the 17th SPIE—Signal Processing, Sensor Fusion, and Target Recognition 6968, Orlando, FL, USA, 16–20 March 2008; pp. 69680F-1–69680F-12.

9.   Amoon, M.; Rezai-rad, G.-A. Automatic target recognition of synthetic aperture radar (SAR) images based on optimal selection of Zernike moments features. *IET Comput. Vis.* **2014**, *8*, 77–85. [CrossRef]

10.   Wright, J.; Yang, A.Y.; Ganesh, A. Robust face recognition via sparse representation. *IEEE Trans. Pattern Anal. Mach. Intell.* **2009**, *31*, 210–227. [CrossRef] [PubMed]

11.   Thiagarajan, J.J.; Ramamurthy, K.N.; Knee, P. Sparse representations for automatic target classification in SAR images. In Proceedings of the IEEE 4th International Symposium on Communications, Control and Signal Processing, Limassol, Cyprus, 3–5 March 2010; pp. 1–4.

12.   Knee, P.; Thiagarajan, J.J.; Ramamurthy, K.N. SAR target classification using sparse representations and spatial pyramids. In Proceedings of the IEEE Radar Conference (RADAR), Kansas City, MO, USA, 23–27 May 2011; pp. 294–298.

13.   Yang, J.; Yu, K.; Gong, Y. Linear spatial pyramid matching using sparse coding for image classification. In Proceedings of the IEEE Conference on Computer Vision and Pattern Recognition, Miami, FL, USA, 20–25 June 2009; pp. 1794–1801.

14.   Lowe, D.G. Distinctive image features from scale-invariant key points. *Int. J. Comput. Vis.* **2004**, *60*, 91–110. [CrossRef]

15.   Dalal, N.; Triggs, B. Histograms of oriented gradients for human detection. In Proceedings of the 2005 IEEE Computer Society Conference on Computer Vision and Pattern Recognition (CVPR'05), San Diego, CA, USA, 20–25 June 2005; pp. 886–893.

16.   Wang, J.; Yang, J.; Yu, K. Locality-constrained linear coding for image classification. In Proceedings of the IEEE Conference on Computer Vision and Pattern Recognition, San Francisco, CA, USA, 13–18 June 2010; pp. 13–18.

17.   Zhang, S.; Sun, F.; Liu, H. Locality-constrained linear coding with spatial pyramid matching for SAR image classification. In *Foundations and Practical Applications of Cognitive Systems and Information Processing, Advances in Intelligent Systems and Computing*; Springer: Berlin/Heidelberg, Germany, 2014; Volume 215, pp. 867–876.

18.   Lai, J.; Jiang, X. Modular weighted global sparse representation for robust face recognition. *IEEE Signal Processing Lett.* **2012**, *19*, 571–574. [CrossRef]

19.   Lazebnik, S.; Schmid, C.; Ponce, J. Beyond bags of features: Spatial pyramid matching for recognizing natural scene categories. In Proceedings of the IEEE Conference on Computer Vision and Pattern Recognition, New York, NY, USA, 17–22 June 2006; pp. 2169–2178.

20.   Lee, H.; Battle, A.; Raina, R.; Ng, A. Efficient sparse coding algorithms. In Proceedings of the Twentieth Annual Conference on Neural Information Processing Systems, Vancouver, BC, Canada, 4–7 December 2006; pp. 801–808.

21.   Needell, D.; Tropp, J.A. CoSaMP: Iterative signal recovery from incomplete and inaccurate samples. *Appl. Comput. Harmon. Anal.* **2008**, *26*, 301–321. [CrossRef]

22.   Mossing, J.C.; Ross, T.D. An evaluation of SAR ATR algorithm performance sensitivity to MSTAR extended operating conditions. *Algorithms Synth. Aperture Radar Imag. V (Proc. SPIE)* **1998**, *3370*, 554–565.

23.   Misha, A.; Mulgrew, B. Radar signal classification using PCA-based features. In Proceedings of the International Conference on Acoustics, Speech, and Signal Processing (ICASSP), Toulouse, France, 14–19 May 2006; pp. 14–19.

24.   Juwei, L.; Kostantions, N.P.; Anastasios, N.V. Face recognition using LDA-based algorithms. *IEEE Trans. Neural Netw.* **2003**, *14*, 195–200. [CrossRef]

Article

Research on Characteristics of Copper Foil Three-Electrode Planar Spark Gap High Voltage Switch Integrated with EFI

Kehua Han [1,2], Wanjun Zhao [1,*], Peng Deng [1], Enyi Chu [2] and Qingjie Jiao [1]

1   State Key Laboratory of Explosion Science and Technology, Beijing Institute of Technology, Beijing 100081, China; kehuahan@126.com (K.H.); 3120195115@bit.edu.cn (P.D.); jqj@bit.edu.cn (Q.J.)
2   National Key Laboratory of Applied Physics and Chemistry, Shaanxi Applied Physics and Chemistry Research Institute, Xi'an 710061, China; 3120195169@bit.edu.cn
*   Correspondence: wanjunzhaowj@bit.edu.cn

Abstract: In view of the low-energy explosion foil detonation system's requirements for the integration technology of high-voltage switches and technical overload resistance technology, a magnetron sputtering coater is used to sputter copper film on the surface of the substrate. The thickness is 4.0 μm, the radius of the main electrode is 4 mm, the trigger electrode is 0.6 mm and 0.8 mm, and the main gaps are 0.8 mm, 1.0 mm, 1.2 mm mm, 1.8 mm, 2.0 mm, 2.2 mm, and 2.6 mm. Copper foil three-electrode planar spark gap high voltage switches are designed and manufactured; and the static self-breakdown characteristics, dynamic operating characteristics, and discharge life characteristics of the three-electrode planar spark gap high voltage switch based on copper foil are studied in this paper. The test results show that with the increase of the main electrode gap from 0.8 mm to 2.6 mm, the self-breakdown voltage of the planar spark gap switch increases, and the working voltage also increases. When the main electrode gap is a maximum of 2.6 mm, the self-breakdown voltage of the switch can reach 3480 V, which indicates that the maximum operating voltage of the switch is 3480 V. When the charging voltage is 2.0 kV, with the increase of the main electrode gap from 0.8 mm to 2.6 mm, the minimum trigger voltage value of the planar spark gap switch increases from 677 V to 1783 V (a = 0.6 mm), and from 685 V to 1766 V (a = 0.8 mm), the switch on time is 16 ns, 22 ns, 28 ns, 48 ns, 64 ns, 77 ns, 93 ns (a = 0.6 mm), and 26 ns, 34 ns, 51 ns, 67 ns, 81 ns, 102 ns (a = 0.6 mm). With the increase of the gap between the two main electrodes of the switch, the maximum static working voltage of the three-electrode plane spark gap high-voltage switch increases, the minimum trigger voltage value also increases, and the on-time of the switch gradually becomes longer. The peak current of the discharge circuit decreases and the dynamic impedance and inductive reactance of the switch also increase; as the width of the trigger electrode increases, the minimum trigger voltage decreases, the dynamic impedance and inductance decrease, and the switch operating voltage with the same parameters is higher. The easier the switch is to turn on, the lower the minimum trigger voltage. The electrode thickness of the three-electrode plane spark gap switch has a certain influence on the field strength and the service life of the switch. The results of this study provide useful references for promoting the research and development of LEEFIs.

Keywords: three-electrode planar spark gap high voltage switch; static characteristics; dynamic characteristics; impedance characteristics; electric field strength

Citation: Han, K.; Zhao, W.; Deng, P.; Chu, E.; Jiao, Q. Research on Characteristics of Copper Foil Three-Electrode Planar Spark Gap High Voltage Switch Integrated with EFI. *Appl. Sci.* **2022**, *12*, 1989. https://doi.org/10.3390/app12041989

Academic Editors: Pingjuan Niu, Li Pei, Yunhui Mei, Hua Bai and Jia Shi

Received: 7 January 2022
Accepted: 6 February 2022
Published: 14 February 2022

1. Introduction

The high-voltage switch is a core key component in exploding foil initiation systems (EFIs). It can not only control the on-off of the current in the initiation discharge circuit, but also has a high turn-off impedance to reduce the power consumption of the exploding foil initiation system: it is more important to have lower on-resistance and inductive reactance to improve the output characteristics of narrow pulse current. Technical indicators and performance parameters directly affect the overall performance of exploding foil initiation systems (EFIs) [1–3]. With the continuous development of low-energy exploding foil initiators (LEEFI), high-voltage discharge switches are also constantly updated. Reynolds Company reported the use of a spark-gap gas discharge switch at the 43rd Annual Meeting of Fuzes in 1999 [4], and the use of an MCT (MOS controlled thyristor) switch was reported at the 45th Annual Meeting of Fuzes in 2001 [5], Xu Cong, Zhu Peng and Chen Kai et al. [6] made Schottky barrier diodes into single-trigger high-voltage switches through MEMS technology in 2017. In 2021, Yang Zhi, Zhu Peng et al. [7] reported the spark-gap switch triggered by a parallel plane sealed on a PCB base. The high-voltage discharge switches reported in the data mainly include three-dimensional cold-cathode-triggered gas-spark-gap high-voltage switches and vacuum-triggered high-voltage switches based on metal ceramic packages, based on IGBT (insulated gate bipolar transistor, insulated gate bipolar transistor) and MCT (metal- oxide-semiconductor controlled thyristor MOS control thyristor) semiconductor high-voltage switches, based on micro-electro-mechanical system (MEMS) and other processes, including the plane electric explosion high-voltage switch and the plane spark gap high-voltage switch.

The cold cathode-triggered gas spark gap high-voltage switch and the vacuum-triggered high-voltage switch based on the three-dimensional structure of the metal-ceramic package have fast closing speed, high operating temperature, small current leakage, and are little influence by radiation. Because this system has good performance and is favored, it has always dominated. However, because it is a ceramic metal package structure, it has the disadvantages of poor mechanical overload resistance, large volume, and high cost. With the continuous development of semiconductor technology, Tom Nickolin [8] proposed a metal oxide semiconductor field effect transistor switch (MOSFET) and N-channel MOS control in LEEFI (low energy exploding foil initiator) at the 45th Fuse Annual Conference in 2001. In 2002, Hanks RL [9] published a design scheme using MCT to control the discharge of high voltage capacitors in a patent. In 2008, Kluge Design Inc (KDI) [10] announced the multiple launch rocket system (MLRS) using LEEFI, indicating that the MCT high voltage switch has achieved engineering applications. In 2016, the U.S. Navy [11] announced a high-voltage switch composed of stacked IGBTs, implemented a discharge test study under short-circuit and load conditions, and planned to use it for live-fire testing at the end of the year. As a typical representative of high-power MOS-controlled semiconductor devices, MCT high-voltage switches have the advantage that conventional MOS gate control signals can be used to control the conduction of the switch, and the pulse current can reach thousands of amperes within 100 nanoseconds, which is suitable for explosion foil ignition with detonation. The on-current of the semiconductor high-voltage switch is large and the working life is long. However, due to its large size, low operating voltage and large leakage current, it is difficult to achieve derating design, resulting in low working reliability and safety of the exploding foil initiation system. In addition, its performance is greatly affected by environmental factors (temperature, electromagnetic, etc.). The above two switches belong to the discrete structure, their volume and the comprehensive performance of the ability to withstand high voltage are low, and the impedance and inductance introduced into the discharge circuit are relatively large, which seriously affects the reliability of the explosion foil initiator system andthe development of miniaturization and low energy of exploding foil initiator.

The planar electric explosion switch is a high-voltage switch that uses the bridge foil of the trigger electrode to generate an electric explosion to generate plasma with conductive properties, thereby making the main electrodes on both sides perpendicular to the trigger electrode conduct. As early as the 1980s, Graham et al. [12] studied the conductive properties of polymer films induced by explosive shock. In 1986, Richardson et al. [13] invented a planar dielectric high-voltage switch suitable for exploding foil initiators (EFI), when the switching dielectric layer is impacted at high speed by the RDX-driven flyer, the upper and lower electrodes of the switch are turned on. In 1989, Nerheim E et al. invented a silicon-based planar electrical explosion high-voltage switch [14], the structure of which is shown in Figure 1. Its manufacturing process is to sequentially deposit switch high-voltage electrodes on a silicon substrate, the trigger electrode is made of amorphous silicon or polysilicon of an electric explosion bridge, and an insulating gap is arranged between the high-voltage electrode and the trigger electrode. Before switching, the two ends of the high-voltage electrode are charged with high voltage, and the trigger electrode is excited by a constant current source, so that when the polysilicon or amorphous silicon bridge foil of the trigger electrode is electrically exploded, a conductive plasma cloud is generated. Under the action of the electric field, the insulating gap between the high-voltage electrodes is broken down and turned on, so as to realize the output of short pulse and large current, and complete the conduction and closing function of the high-voltage switch.

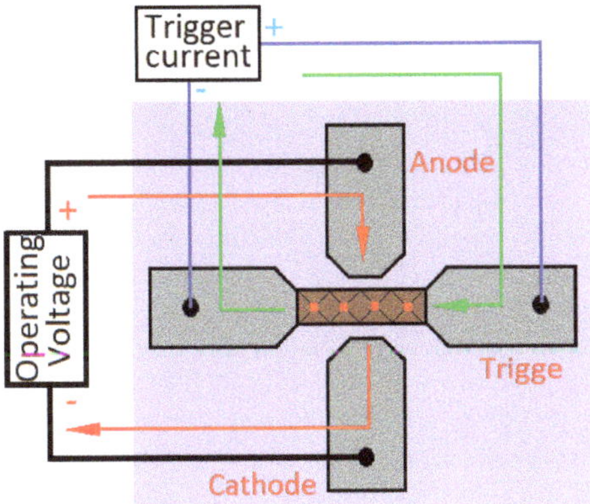

Figure 1. Schematic representation of planar-electric explosion switch based on silicon substrate.

In 2009, Baginski T A et al. [15,16] designed and manufactured a planar dielectric explosion high-voltage switch triggered by Schottky diode and a micro-bridge explosion planar switch with a series structure, and proposed the idea of integrating the switch with EFI. The switch structure is shown in Figure 2.

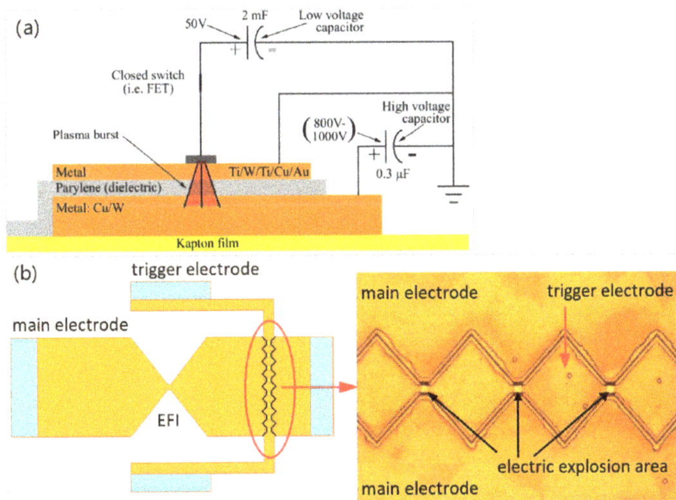

Figure 2. Schematic representation of planar high-voltage switch with SBD trigger and planar-electric explosion switch with series structure. (**a**) Planar high-voltage switch with SBD trigger. (**b**) Planar-electric explosion switch with series structure.

The planar dielectric explosion high voltage switch based on Schottky diode is composed of substrate, lower electrode, dielectric layer, upper electrode, and Schottky diode. By applying a reverse voltage to the Schottky diode to cause reverse breakdown, and then under the thermal effect of pulse current, an electrical explosion is caused, resulting in dielectric breakdown; the upper and lower electrodes are connected, and the switch is turned on. The series-structured micro-bridge explosive planar switch is based on the silicon-based planar electro-explosive high-voltage switch, and adopts the series connection of multiple planar electro-explosive high-voltage switches to improve the electrical performance and reliability of the planar electro-explosive switch. In 2011, Baginski T A et al. [17] proposed a planar trigger switch (PTS) containing a polyimide film insulating layer. The current test and simulation study of the switch discharge circuit are completed, and the test results are good. Combined with LEEFI, the HNS-IV was successfully detonated, verifying the practicability of the switch. In 2012, Zhou Mi, Han Kehua et al. [18] prepared a single-bridge planar electric explosion switch based on copper film by ion etching method, and studied the effect of gap distance on the switch performance. The results show that as the gap distance decreases, the action time of the switch decreases. In 2018, Wang Runyu et al. [19] used an improved manufacturing process to replace the wet etching process to prepare a miniature metal bridge foil explosive planar switch, and studied the influence of the thickness of the adhesive layer and the insulation method on the insulation effect in the switch insulation treatment. In 2020, Xu Cong et al. designed three trigger modes: the Schottky diode [20], pn junction diode [21], and micro bridge foil [22,23] based on the planar dielectric explosion high voltage switch of the Schottky diode, such as the Baginski TA flat dielectric high voltage switch. The electrical characteristics of the three switches are preliminarily studied, and the results show that, among the three trigger mode switches, the micro-foil planar dielectric switch can obtain the highest peak current and the shortest rise time at a lower operating voltage. When the planar electric explosion high-voltage switch is turned on and closed, a trigger voltage needs to be added to make the core part of the switch generate an electric explosion instantaneously and complete the closing function of the switch, indicating that this type of switch can only perform a single action. Because the characteristics of the planar electric explosion high-voltage switch is a one-time function, the conduction of the switch is not reversible, so the switch cannot complete the testability before the system is used in the exploding foil initiation system, which seriously affects the reliability and safety of the system.

The use of micro-electromechanical machining technology to planarize the spark gap high-voltage switch can effectively solve the above problems. The planar spark gap high-voltage switch can not only complete multiple discharge functions, but also improve the switch's resistance to mechanical overload, reduce costs, and reduce system volume. It can also realize the integration function of the high-voltage switch and the explosion foil, reduce the parasitic impedance and inductive reactance in the discharge circuit, reduce the energy consumption of the system, and improve the integration degree of the system. In this paper, a copper foil-based three-electrode planar spark-gap high-voltage switch is designed and fabricated by using a magnetron sputtering coater to sputter copper film on the surface of the substrate. The static self-breakdown characteristics, dynamic operating characteristics, and discharge life characteristics of the three-electrode planar spark gap high voltage switch based on copper foil are studied.

2. Design and Fabrication of a Three-Electrode Planar Spark Gap High-Voltage Switch Based on Copper Foil

2.1. The Design of Switches

The three-electrode planar spark gap high voltage switch was composed of two main electrodes and trigger. The main electrode includes the cathode and anode, as shown in Figure 1. The main electrode is semicircular in shape and placed in the opposite position. The trigger electrode is located in the middle of the two main electrodes. Its principle is to load high DC voltage between the cathode and anode of the switch. When a specific pulse voltage signal is applied to the trigger electrode, a high voltage gap is formed between the cathode and the trigger electrode. A certain number of ions or electrons are produced by the breakdown field strength, and the particles and gas undergo the collision multiplication process, resulting in the instantaneous conduction of the anode and cathode of the switch. In Figure 3, *a* represents the width of the trigger electrode, *b* represents the gap width of the main electrode, and *R* represents the radius of the main electrode.

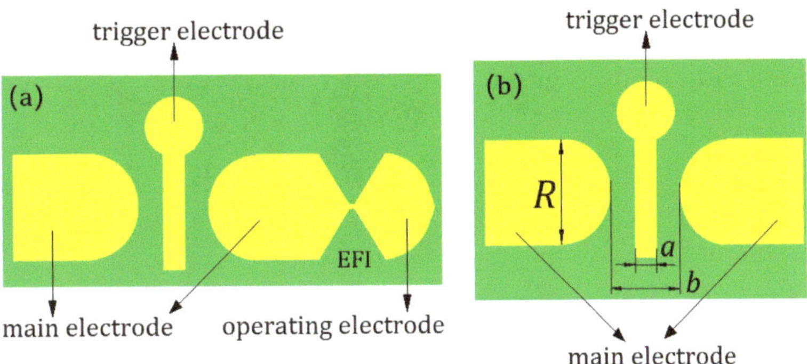

Figure 3. Integrated exploding foil initiator based on three-electrode planar spark gap high-voltage switch. (**a**) Copper foil three-electrode planar spark-gap high-voltage switch integrated with EFI. (**b**) Three-electrode planar spark-gap high-voltage switch based on copper foil.

As can be seen from Figure 3, the key structures of the planar three-electrode switch include the main electrode radius (R), the main gap (b), the trigger gap ($(b - a)/2$) and the trigger electrode width (a). The diameter R of the two main electrodes is a semi-circular structure with a diameter of 4.0 mm $\pm$ 0.5 mm. The purpose of this design is to ensure that a uniform electric field exists between the electrode gaps as much as possible, so as to improve the service life of the switch. With the input of the trigger signal, the switch conduction first occurs in the trigger gap. Therefore, reducing the trigger gap is beneficial to improve the working reliability of the switch, but the gap is too small, which affects the working voltage range and safety of the switch. Combined with the machining accuracy, the structural design parameters for the plane three-electrode spark gap high-voltage switch are shown in Table 1.

Table 1. Switch Structure Parameters.

Structure Name	Code	Parameter
Switch thickness	d	4 μm $\pm$ 0.1 μm
main electrode radius	R	4.0 mm $\pm$ 0.5 mm
main gap	b	0.8/1.0/1.2/1.8/2.0/2.2/2.6 mm
Trigger electrode width	a	0.6/0.8 mm
Trigger gap	$(b - a)/2$	0.1/0.2/0.3/0.6/0.7/0.8/1.0 mm 0.1/0.2/0.5/0.6/0.7/0.9 mm

2.2. Switch Fabrication and Characterization

In the three-electrode planar spark gap high voltage switch, a magnetron sputtering coater was used to sputter copper film on the surface of the substrate. The coating photoresist is spined on the surface of the copper film. Then the photoresist surface is covered with a photoresist mask, which is exposed under a strong light to develop the substrate. Finally, the final switch is formed after being etched with $FeCl_3$ etching solution. The switch is shown in Figure 4.

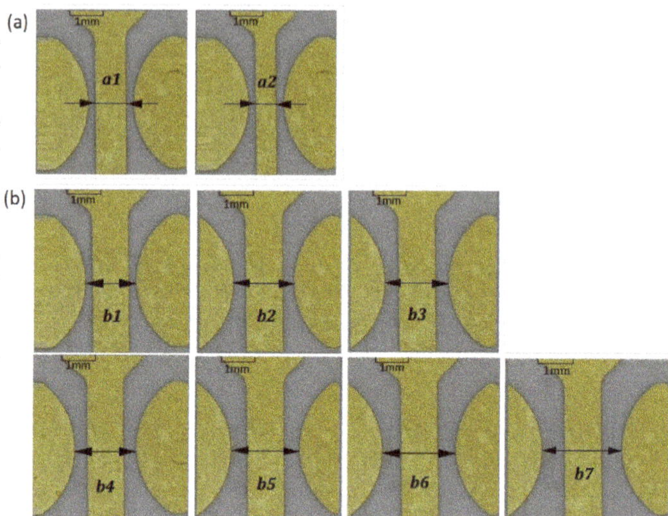

Figure 4. Physical diagram of three-electrode plane spark gap high-voltage switch. (**a**) Planar switching with different trigger electrodes. (**b**) Planar switching with different main electrode gaps.

The microscope stage micrometer is used to test the switch structure size according to the parameter requirements in Table 1. The test results are all within the design range requirements.

3. Research on the Characteristics of High-Voltage Switching with Three-Electrode Plane Spark Gap

3.1. Test Device and Principle

For the performance test of the three-electrode planar spark gap high-voltage switch, the test device includes a high-voltage power supplied with the output voltage of 0~4 kV and accuracy of 1‰, a pulse trigger power supplied with the adjustable output voltage of 0~3 kV, a rising pulse time of no more than 100 ns and falling pulse time of no more than 1 µs, a digital high-voltage meter with input impedance of 1000 mΩ and measurement accuracy of 1‰, the Rogofiski Roche current measuring coil with the model specification of 5008c, a voltage divider of 1/1000, and a high voltage pulse capacitor with the model specification of C471/0.2 µF/3.0 kV. The test principle is shown in Figure 5.

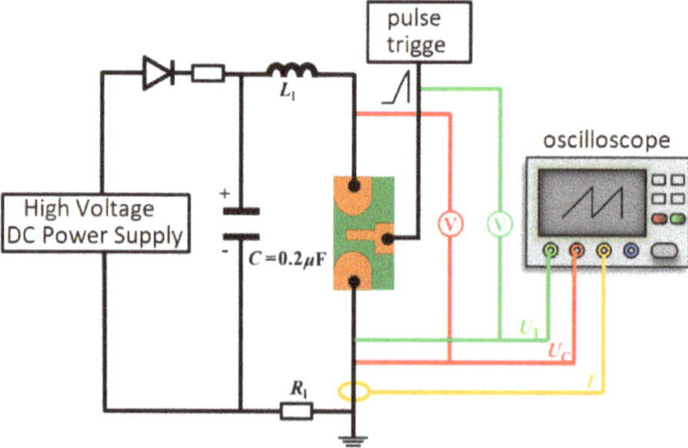

Figure 5. Schematic diagram of test principle.

As shown in Figure 5, the high-voltage DC power supply outputs the working voltage U_{SB} to the positive electrode of the high-voltage capacitor and the three-electrode plane spark gap high-voltage switch, and the pulse trigger power supply provides the trigger voltage $U_{trigger}$ to the three-electrode plane spark-gap high-voltage switch. When the electric field strength between the trigger electrode and the main electrode is greater than the breakdown strength of the air in the gap, a breakdown occurs between the gaps, the electric field in the gap is distorted, and a large number of electrons in the negative electrode of the main electrode rapidly move to the positive electrode to turn on the switch. The high-voltage capacitor discharges the load through the three-electrode plane spark gap high-voltage switch, and uses the Rogowski coil, the high-voltage probe and the voltage divider, and the oscilloscope to test the discharge current, working voltage, and trigger voltage in the loop, respectively.

3.2. Static Self-Breakdown Characteristics

The three-electrode planar spark gap high-voltage switch based on copper foil, like other high-voltage switches, needs to have a specific operating voltage range; that is, the gap between the main electrode and the trigger electrode of the switch can withstand a certain high voltage without self-breakdown. Therefore, the self-breakdown voltage U_{SB} of the switch is also the maximum working voltage of the switch. The high voltage capacitor (0.2 μF) is slowly charged at a rate of about 20 V/s from the high voltage DC power supply, until the air in the gap between the two electrodes occurs self-breakdown. A high-voltage probe and Rogowski coil were used to record the self-breakdown voltage U_{SB} and loop current I changes. When the width a of the trigger electrode is 0.6 mm and the width b of the gap between the two main electrodes is 1.2 mm, the test waveform of the self-breakdown characteristic of the switch is shown in Figure 6a. The maximum operating voltage U_{SB} is 2361 V, and the maximum output current peak value is 2892 A. The U_{SB} test curves of different parameters of the plane spark gap switch by the orthogonal method are shown in Figure 6b.

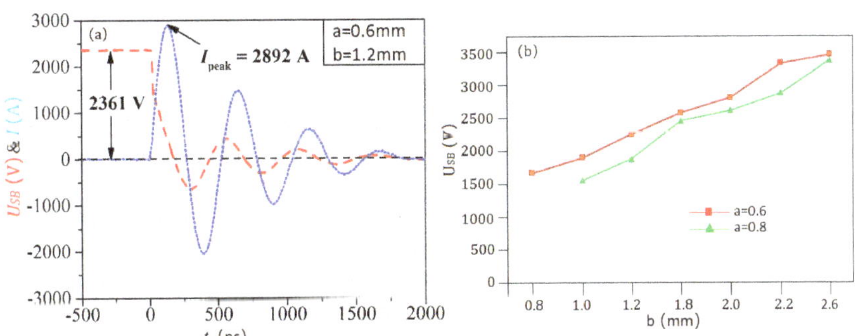

Figure 6. Waveform and curve of self-breakdown characteristic of three-electrode planar switch. (**a**) Self-breakdown characteristic waveform (C = 0.2 μF) Figure. (**b**) Self-breakdown characteristic curve.

It can be seen from Figure 6.b that as the main electrode gap increases from 0.8 mm to 2.6 mm, the self-breakdown voltage of the planar spark gap switch gradually increases, and the operating voltage also increases. When the main electrode gap is a maximum of 2.6 mm, the self-breakdown voltage of the switch can reach 3480 V, which indicates that the maximum operating voltage of the switch is 3480 V. According to Thomson's theory and Baschen's law [24,25] of uniform electric field self-discharge under low pressure, the breakdown voltage U_{SB} of the air gap is related to the air pressure p and the main electrode gap b, and under the condition of a fixed gas atmosphere, the maximum operating voltage U_{SB} is positively correlated with b. By carrying out a large number of self-breakdown characteristic tests, it is shown that the self-breakdown voltage fluctuates greatly (the range is about 180 V), and the larger the gap, the greater the fluctuation. According to the principle of electrostatic discharge, this is because the trigger electrode is right-angled, and tip discharge is prone to occur during the conduction process, which makes the field strength here larger, resulting in an increase in the unevenness of the electric field.

3.3. Dynamic Operating Characteristics

3.3.1. Dynamic Minimum Trigger Voltage

In order to determine the conduction condition of the three-electrode planar spark gap high voltage switch with seven kinds of main electrode gaps, the three-electrode planar spark gap high voltage switch with main electrode gaps of 0.8 mm, 1.0 mm, 1.2 mm, 1.8 mm, 2.0 mm, 2.2 mm, and 2.6 mm are designed, and the trigger electrode widths are 0.6 mm and 0.8 mm, respectively. With the charging voltage of 2.0 kV, the minimum trigger voltage curves of various switch parameters are shown in Figure 7.

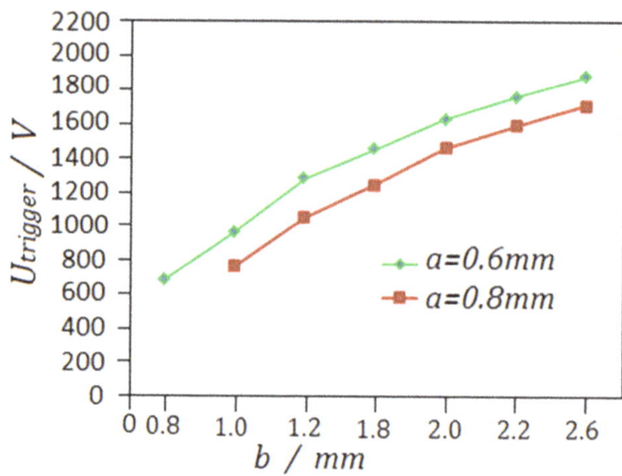

Figure 7. The curves of minimum trigger voltage varying with gap.

As shown in Figure 7, under the condition of charging voltage of 2.0 kV, as the main electrode gap increases from 0.8 mm to 2.6 mm, the minimum trigger voltage value of the planar spark gap switch increases from 677 V to 1783 V (a = 0.6 mm), 685 V rises to 1766 V (a = 0.8 mm). This shows that with the continuous increase of the main electrode gap, the minimum trigger voltage of the switch increases. With the same gap, the width of the trigger electrode is wider, the minimum trigger voltage becomes lower, for the reason that the conduction principle of the three-electrode planar spark gap high voltage switch is to apply pulse trigger voltage to the trigger electrode. Herein, the gap electric field is distorted and the air breakdown effect occurs between the gaps, making the two poles of the switch conduct. When the three-electrode planar spark gap high voltage switch is in the triggering state, the average electric field strength between the two poles is calculated as follows [26,27]:

$$E = \frac{U_{SB}}{b - a} \tag{1}$$

E: Average electric field, V/m;
U_{SB}: Working voltage, V;
b: Gap, mm;
a: Trigger electrode width.

As shown in Formula (2), under the condition of the same trigger electrode width, when the applied voltage U_{SB} between the two poles is constant, the average electric field strength E increases with the decrease of gap b, meaning that the smaller the gap between the two poles of the switch is, the greater the average electric field strength becomes, and the easier the air gas breakdown effect occurs, so the energy required for the trigger electrode is lower. On the contrary, the average electric field strength between the two poles of the switch is stronger and the energy required for the trigger is higher, when the switch is on. As shown in Formula (2), under the condition of the same gap, when the applied voltage U_{SB} between the two poles is fixed, the width of the trigger electrode is increased, the average electric field strength increases, and the air breakdown effect occurs easily, resulting in the lower energy required for the trigger electrode.

The matching test between the working voltage and the trigger voltage was carried out. The test results show that the relationship between the trigger voltage and the working voltage is inversely proportional. When the working voltage was high, the switch turned on easily, and the minimum trigger voltage was reduced. That is because when a relatively high voltage is applied between the two poles of the switch, the average electric field strength increases, and it is easier for breakdown effect to take place between the two poles of the switches, the trigger pole and air interface. Therefore, the required trigger energy will become low. The fitting curve of the relationship between the working voltage of the switch and the trigger voltage was shown in Figure 8.

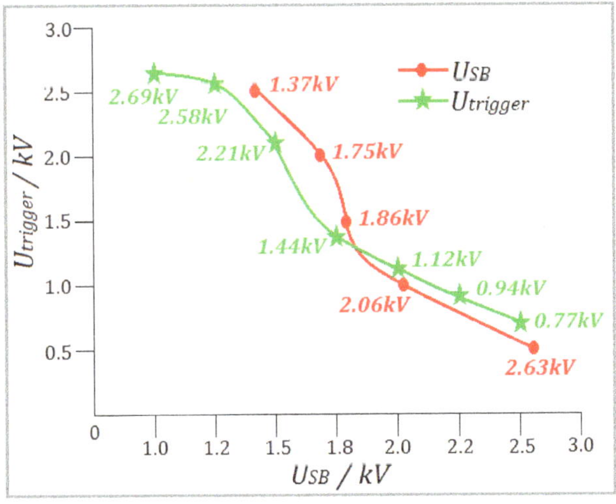

Figure 8. Relation curves between trigger voltage and working voltage.

3.3.2. Switch Dynamic Conduction Performance

The time t_{on} is the conduction time of the three-electrode planar spark gap high voltage switch. This is measured from the time when applying the trigger pulse voltage for the trigger electrode to the time when completely connection of the two poles of the switch and the oscillation current occurred in the discharge circuit. When the trigger electrodes are 0.6 mm and 0.8 mm, the trigger voltage is 1.8 kV, and the working voltage is 2.0 kV, the switch on-time t_{on} test results of different gaps are 16 ns, 22 ns, 28 ns, 48 ns, 64 ns, 77 ns, and 93 ns (a = 0.6 mm) and 26 ns, 34 ns, 51 ns, 67 ns, 81 ns, and 102 ns (a = 0.8 mm). The test waveforms are shown in Figure 9.

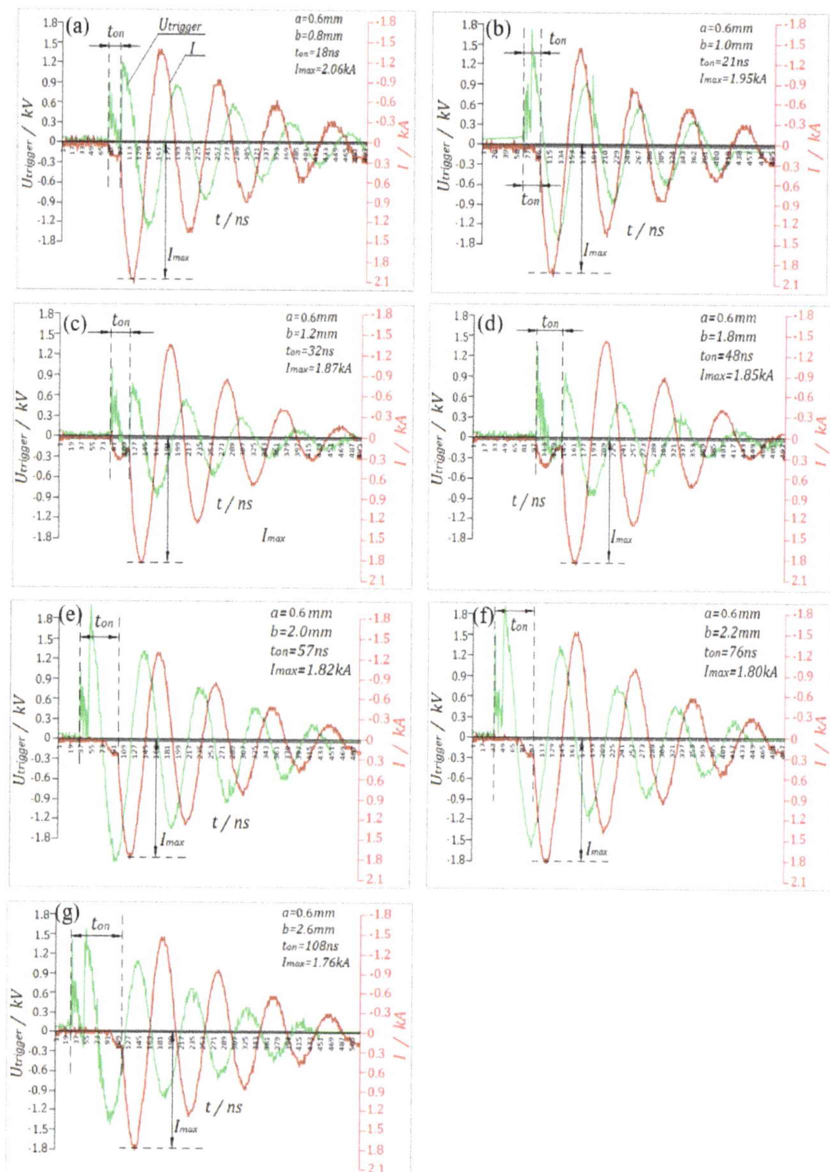

Figure 9. The tested waveform of switch conduction performance (**a**) b = 0.8 mm, (**b**) b = 1.0 mm, (**c**) b = 1.2 mm, (**d**) b = 1.8 mm, (**e**) b = 2.0 mm, (**f**) b = 2.2 mm, (**g**) b = 2.6 mm.

In Figure 9, the tested waveform of the on-off performance of the switches can be seen. With the increasing gap, the on-off time of the switch becomes gradually longer, and the peak current of the discharge circuit is reduced. When the switch is on, the breakdown effect starts between the pulse trigger voltage and the strong electric field with the air interface, resulting in the electric field distortion between the two poles of the switches. A certain number of ions or electrons are generated instantaneously and move with a high speed under the electric field of the two poles of the switch. When the working voltage, trigger voltage, and the width of trigger electrode are fixed, with the increase of the switch gap, the average electric field strength between the two poles of the switch decreases, which leads to the decrease of the velocity of ions or electrons generated when the switch is on, so that the on-time of the switch becomes longer. On the contrary, when the switch gap is low, with the increase of the average electric field, the average electric field strength between the two poles of the switch decreases, the velocity of ions or electrons increases and the on-time decreases. In addition, with the increase of the gap between the two poles of the switch, the average electric field strength between the two poles of the switch decreases. When the breakdown effect occurs between the trigger pole and the air interface, the energy loss is large, leading to a decrease in the peak current in the circuit. The regular curves between the trigger voltage and operating voltage of the switch and the on-time of the switch are shown in Figure 10.

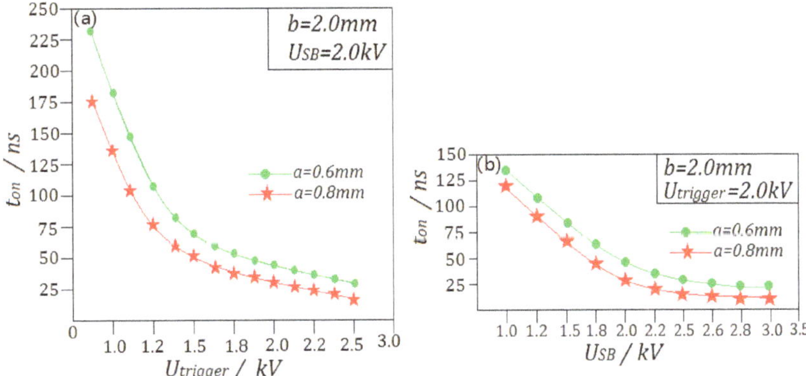

Figure 10. Curve of factors affecting the conduction time of the three-electrode plane spark gap high-voltage switch. (**a**) Trigger voltage and on-time relationship curve. (**b**) Working voltage and on-time relationship curve.

3.3.3. Switch Dynamic Impedance and Inductive Reactance Characteristics

In order to realize the function of switching and disconnecting the narrow pulse current in the exploding foil initiation circuit, the high voltage switch not only has a higher turn-off impedance to reduce the power consumption of the exploding foil initiation system, but also has a lower conductive impedance and inductive reactance to improve the narrow pulse current output characteristics of the high-voltage pulse capacitor.

In principle, the initiation circuit of exploding foil can be equivalent to a RLC series circuit [28–30]. The parasitic resistance and inductance of the three-electrode planar spark gap high voltage switch can be calculated by measuring the waveform parameters of discharge oscillation current of the circuit with oscilloscope.

The formula of parasitic inductance is as follows.

$$L_0 = L - l \tag{2}$$

$$L = \frac{\overline{T}^2}{4\pi^2 C} \tag{3}$$

The formula of parasitic resistance is as follows.

$$R_0 = R - r \tag{4}$$

$$R = \frac{2L}{\overline{T}} \ln \xi, \text{ among them, } \xi = \frac{\sum\limits_{j=1}^{n} \lambda_j}{n}, \lambda_j = \frac{I_j}{I_{j+1}} \tag{5}$$

L: Total inductance of discharge circuit, nH;
l: Load inductance, nH;

R: Total resistance of discharge circuit, mΩ;
r: Load resistance, mΩ;
$\overline{T}$: The average period of oscillation, µs;
I_{j+1}, I_j: The value of forward oscillation current, kA;
ξ: The average coefficient of current attenuation;
C: High voltage pulse capacitance, µF;

According to the test and calculation results, the corresponding data of the relationship between the gaps of two main electrodes, the width of trigger electrode, the dynamic impedance, and inductive reactance of the switch are shown in Tables 2 and 3, and the corresponding curves are shown in Figure 11.

Table 2. The conductive impedance data of the switch.

b/mm R_0/mΩ a/mm	0.8	1.0	1.2	1.8	2.0	2.2	2.6
0.6	32.8	51.2	88.4	117.2	133.1	171.8	223.9
0.8	/	42.7	68.5	96.7	127.0	153.9	207.1

Table 3. The inductive reactance data of conductive switch.

b/mm L_0/nH a/mm	0.8	1.0	1.2	1.8	2.0	2.2	2.6
0.6	19.3	21.8	29.4	33.7	38.2	42.6	45.3
0.8	/	16.6	19.1	23.5	26.2	30.2	37.6

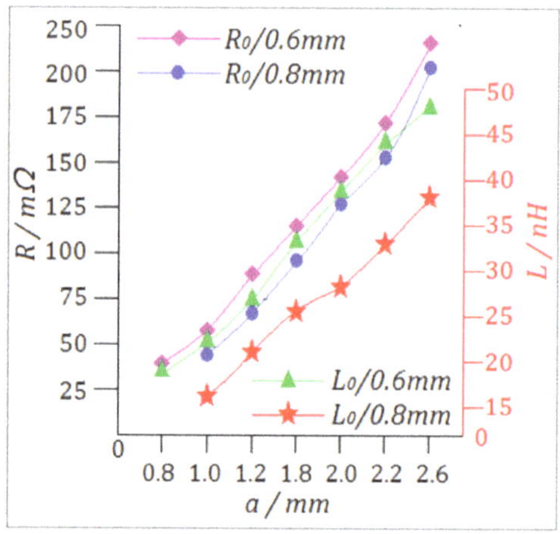

Figure 11. Corresponding curves of switch parameters with dynamic impedance and inductive reactance of the switch.

As shown in Figure 11 and Tables 2 and 3, with the increase of the gap between the two main electrodes of the switches, the dynamic impedance and inductive reactance of the switches is increased, but they will decrease with the increase of the width of the trigger electrode. The increase range of impedance relativity is larger than that of the inductive reactance. With the increase of the gap between the two main electrodes of the switch, the average electric field strength between the two main electrodes had been reduced and the concentration of ions or electrons had been decreased, resulting in the decrease of electric current density passing through. Therefore, the impedance and inductive reactance will increase. As the width of trigger electrode increased, the gap between the two main electrodes becomes smaller and the average electric field strength between two main electrodes increases, resulting in lower impedance and inductive reactance.

3.4. Discharge Life Characteristics of Planar Spark Gap Switch

The service life of the high-voltage switch seriously affects the working reliability of the exploding foil initiation system. The performance of the three-electrode planar spark-gap high-voltage switch based on copper foil was reduced after several discharge tests. Under the same input voltage condition, the working reliability is reduced, the conduction time is longer, and the trigger voltage needs to be increased. Moreover, it was found that the trigger electrode of the switch had traces of burns, as shown in Figure 12.

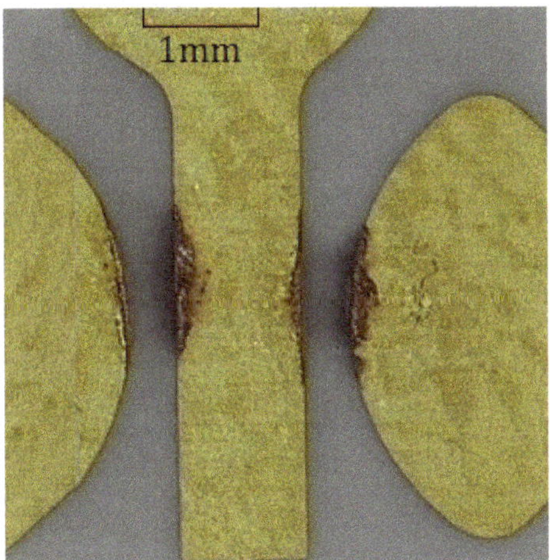

Figure 12. Photo of ablation after switch discharge.

After analysis, this is because the electrode gap is filled with air, so that the three-electrode planar spark gap switch can withstand a specific high voltage, but when the switch needs to be triggered and turned on, it needs to form a sufficient electric field strength in the trigger gap. Since the thickness of the copper foil-based three-electrode plane spark gap high-voltage switch is only 4.0 µm, a large amount of ions or electrons will be generated in the instant when the switch is turned on, and the narrow pulse strong current generated by the discharge circuit will be turned on, causing burns to the switch electrodes.

The electrode thickness of the three-electrode plane spark gap switch has a certain influence on the field strength, and ultimately affects the service life of the switch. The electrostatic field of the three-electrode plane switch is analyzed with the help of finite element simulation software. In the simulation, copper was selected as the electrode material, and air was selected as the dielectric material. The excitation source was electrostatic field solver excitation source, and the boundary condition is balloon boundary condition. The diameter of positive and negative main electrodes is 4.0 mm, the distance b is 2.0 mm and the width of trigger electrode a is 1.2 mm. The setting negative voltage is 0 V, the positive voltage is 1.3 kV, the trigger voltage is 1.5 kV, the number of calculation steps is 10, and the allowable error is 0.1%. Figure 13 shows the overall field intensity distribution cloud diagram of the three-electrode plane spark gap high voltage switch, Figure 14 shows the electric field distribution before triggering, and Figure 15 shows the electric field distribution after triggering.

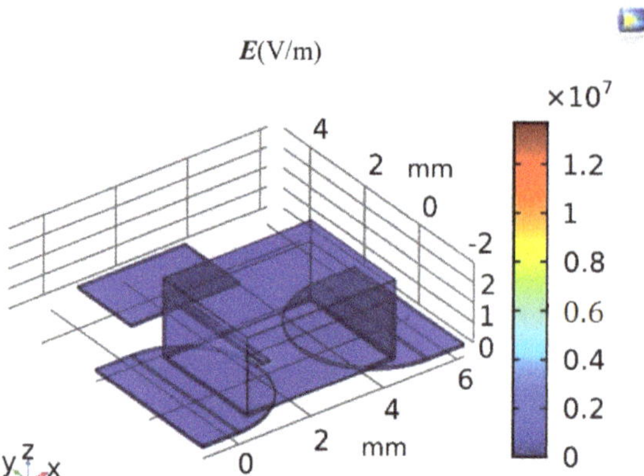

Figure 13. Cloud diagram of the overall field intensity distribution of the three-electrode plane spark gap high-voltage switch.

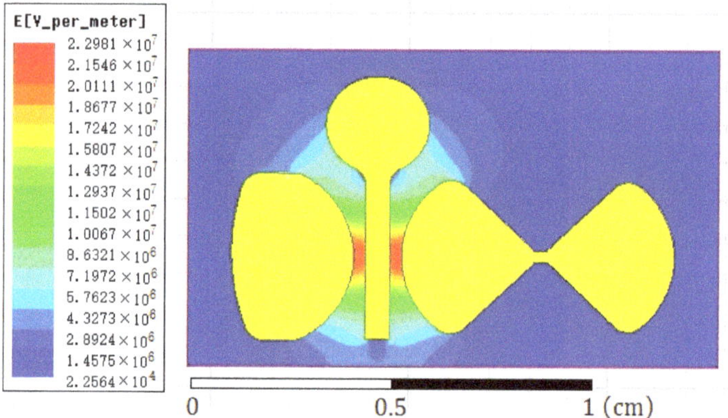

Figure 14. Electric field distribution before triggering.

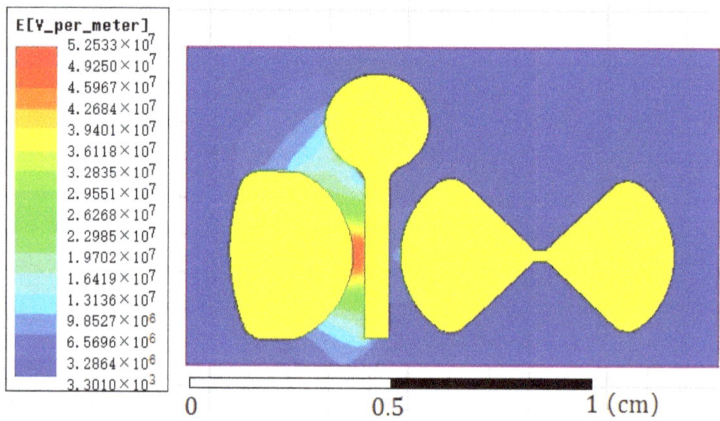

Figure 15. Electric field distribution after triggering.

As shown in Figure 14, when a high voltage of 1.3 kV is provided between the two main electrodes of the switch, the electric field is evenly distributed between the two main electrodes, and the field strength near the edge of the trigger electrode is the largest at 22.9 kV/cm. At this time, the maximum field strength is lower than the breakdown strength of air by 30 kV/cm, so the switch cannot self-breakdown. Considering that the air in the test environment is a non-ideal environment, the breakdown strength of the air will be lower than the ideal 30 kV/cm due to factors such as humidity and temperature. In order to improve the safety of the switch, it is necessary to reserve a certain distance to ensure the insulation effect and avoid false triggering. As shown in Figure 15, the trigger electrode was applied with a voltage of 1.5 kV, and the maximum electric field strength between the trigger electrode and the cathode reached 52.5 kV/cm, which was 30 kV/cm higher than the breakdown strength of air. At this time, the gas between the trigger electrode and the cathode will be ionized, and electrical breakdown will occur, so that the main electrodes will be broken down, and the conduction loop will discharge. In order to improve the working reliability of the switch, when the trigger voltage is loaded, the minimum electric field strength between the two main electrodes and the trigger electrode must be higher than the breakdown strength of air by 30 kV/cm.

The maximum static operating voltage of the three-electrode planar switch is simulated by the established simulation model. In the simulation, the dielectric material is air, no trigger voltage is applied, the negative voltage is 0 V, and the anode voltage constantly increases from 0 V. When the electric field strength of the main electrode gap is greater than the air breakdown strength, the switch is considered to be broken down. Since the breakdown strength of ideal air is 30 kV/cm, the lower limit of the field strength is set to 30 kV/cm in the simulation results; only the electric field strength greater than the breakdown strength of air is displayed, so that the self-breakdown process of the switch can be visually observed, and the self-breakdown voltage can be obtained, as shown in Figure 16.

Figure 17 is a partial enlarged view of the field strength of the switch gap. It can be seen from Figure 17 that the field strength of the trigger gap has increased significantly, and the field strength is greater as it is closer to the trigger electrode. The electrostatic field simulation results show that the discharge first occurs in the trigger gap, and then extends to the entire main electrode gap, but the field strength is the highest around the main electrode closest to the trigger electrode, and a high concentration of electron clouds appears. However, since the thickness of the switch electrode is only 4 μm, a burning phenomenon occurs.

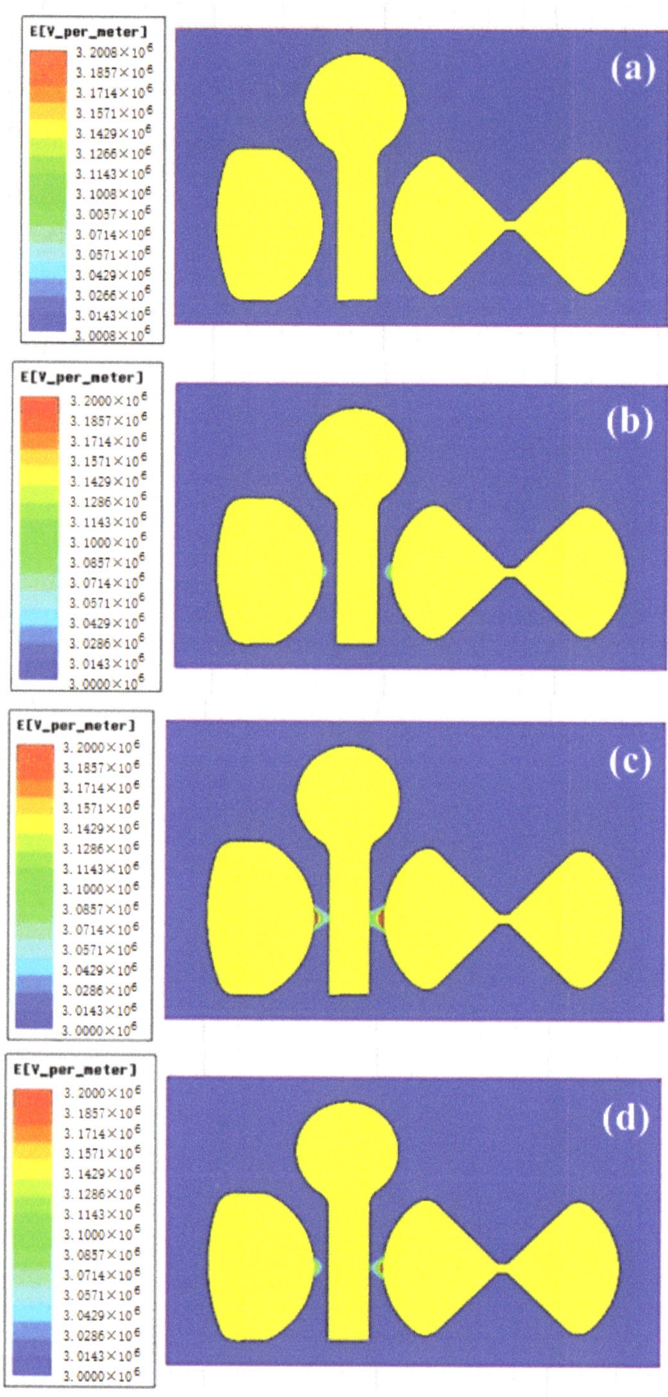

Figure 16. Electric field distribution of integrated components under different charging voltages (**a**) electric field distribution at 1500 V, (**b**) electric field distribution at 1500 V, (**c**) electric field distribution at 2500 V, (**d**) electric field distribution at 2500 V, (**d**) Electric field distribution at 2558 V.

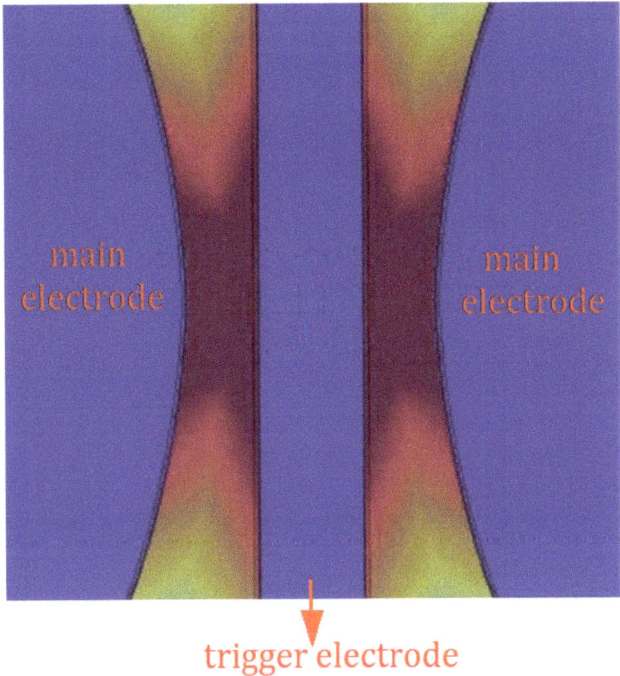

Figure 17. Simulation results of the field strength distribution in the trigger gap of the three-electrode plane spark gap high-voltage switch.

4. Conclusions

In view of the low-energy explosion foil detonation system's requirements for the integration technology of high-voltage switches and the technical overload resistance technology, a magnetron sputtering coater is used to sputter copper film on the surface of the substrate. The thickness is 4.0 μm, the radius of the main electrode is 4 mm, the trigger electrode is 0.6 mm and 0.8 mm, and the main gaps are 0.8 mm, 1.0 mm, 1.2 mm mm, 1.8 mm, 2.0 mm, 2.2 mm, and 2.6 mm. Copper foil three-electrode planar spark gap high voltage switches are designed and manufactured, and the static self-breakdown characteristics, dynamic operating characteristics, and discharge life characteristics of the three-electrode planar spark gap high voltage switch based on copper foil are studied in this paper. The test results show that with the increase of the main electrode gap from 0.8 mm to 2.6 mm, the self-breakdown voltage of the planar spark gap switch increases, and the working voltage also increases. When the main electrode gap is a maximum of 2.6 mm, the self-breakdown voltage of the switch can reach 3480 V, which indicates that the maximum operating voltage of the switch is 3480 V. Under the condition of charging voltage of 2.0 kV, with the increase of the main electrode gap from 0.8 mm to 2.6 mm, the minimum trigger voltage value of the planar spark gap switch increases from 677 V to 1783 V (a = 0.6 mm), and from 685 V to 1766 V (a = 0.8 mm), the switch-on times are 16 ns, 22 ns, 28 ns, 48 ns, 64 ns, 77 ns, 93 ns (a = 0.6 mm), and 26 ns, 34 ns, 51 ns, 67 ns, 81 ns, 102 ns (a = 0.6 mm). With the increase of the main electrode gap, the maximum static operating voltage of the three-electrode planar spark gap high voltage switch increased. When the same width of trigger electrode was used, the minimum trigger voltage increased, with the increase of the main electrode gap. When the same width of the trigger electrode was used, the minimum trigger voltage decreased. When the switch had the same parameters, the trigger voltage was inversely proportional to the working voltage. When the same width of the trigger electrode was used, with the increase of the gap between main electrodes, the conduction time of the switch was longer, and the peak current of the discharging circuit decreased. The dynamic impedance and inductive reactance of the switch increase with the increase of the gap between the two main electrodes, and decrease with the increase of the width of the trigger electrode.

Author Contributions: Conceptualization, K.H., Q.J. and E.C.; methodology, K.H. and W.Z.; software, P.D. and K.H.; validation, K.H. and W.Z.; formal analysis, W.Z. and Q.J.; investigation, K.H., Q.J. and E.C.; resources, E.C. and K.H.; data curation, K.H., W.Z. and Q.J.; writing—original draft preparation, K.H., W.Z. and Q.J.; writing—review and editing, K.H., W.Z. and Q.J.; visualization, P.D. and K.H.; supervision, Q.J. and E.C.; project administration, Q.J. and E.C.; funding acquisition, K.H. and W.Z. All authors have read and agreed to the published version of the manuscript.

Funding: We appreciate the funding support from the National Natural Science Foundation of China (Grant No. 22105025 and Grant No. 52022013). Thanks for the support from the China Postdoctoral Science Foundation (Grant No. 2021M690376).

Institutional Review Board Statement: Not applicable.

Informed Consent Statement: Not applicable.

Data Availability Statement: Not applicable.

Conflicts of Interest: The authors declare no conflict of interest. The founding sponsors had no role in the design of the study; in the collection, analyses, or interpretation of data; in the writing of the manuscript, and in the decision to publish the results.

References

1. Strand, T.; Berzins, L.V.; Goosmanl, D.R. Velocimetry using heterodyne techniques. In Proceedings of the 26th International Congress on High-Speed Photography and Photonics, Alexandria, VA, USA, 20–24 September 2004; pp. 593–599.
2. Prinse, W.; Scholtes, G. A Development Platform for a Microchip EFI. In Proceedings of the 52nd Annual Fuze Conference, Sparks, NV, USA, 13–15 May 2008.
3. Han, K.-h.; Zhou, J.; Rec, X.; Liu, T.; Ao, C.-g.; Tong, H.-h. Effect of High Voltage Pulse Power Source Equivalent Parameter on Exploding Performance of Foil Bridge. *Hanneng Cailiao/Chin. J. Energ. Mater.* **2014**, *22*, 828–833.
4. Chu, K.W.; Scott, G.L. *A Comparison of High-Voltage Switches*; Sandia National Labs.: Albuquerque, NM, USA; Livermore, CA, USA, 1999.
5. Jim, R.D. Fireset for a Low Energy Exploding Foil Initiator: SCR Drive MOSFET Switch. US Patent 6138571, 31 October 2000.
6. Xu, C.; Zhu, P.; Chen, K.; Zhang, W.; Shen, R.; Ye, Y. A highly integrated conjoined single shot switch and exploding foil initiator chip based on MEMS technology. *IEEE Electron Device Lett.* **2017**, *38*, 1610–1613. [CrossRef]
7. Yang, Z.; Wang, K.; Zhu, P.; Liu, P.; Zhang, Q.; Xu, C.; Jian, H.T.; Shen, R.Q. A reusable planar triggered spark-gap switch batched-fabricated with PCB technology for medium-and low-voltage pulse power systems. *Defence Technol.* **2021**, *17*, 1572–1578. [CrossRef]
8. Fujii, K.; Koellensperger, P.; Doncker, R. Characterization and comparison of high blocking voltage IGBTs and IEGTs under hard- and soft-switching conditions. *IEEE Trans. Power Electron.* **2008**, *23*, 172–179. [CrossRef]
9. Hanks, R.L. Low-Jitter High-Power Thyristor Array Pulse Driver and Generator. US Patent 6462605 B1, 8 October 2002.
10. Salyers, P. Guided MLRS electronic safety & arming devices (ESAD) & electronic safety& arming fuze (ESAF). In Proceedings of the 43rd Gun & Missile Conference, NAIA, New Orleans, LA, USA, 21–24 April 2008.
11. Soto, G. Navy Overview. In Proceedings of the 59th Annual Fuze Conference, Charleston, SC, USA, 4 May 2016.
12. Graham, R.A. Shock-induced electrical switching in polymeric films. In *Mega-Gauss Physics and Technology*; Springer: New York, NY, USA, 1980.
13. Richardson, D.D. *The Effect of Switch Resistance on the Ring down of a Slapper Detonator Fireset*; MRL-R1004; Materials Research Laboratories: Struthers, OH, USA, 1986.
14. Nerheim, E. Integrated Silicon Plasma Switch. US Patent 4840122, 20 June 1989.
15. Han, K.; Deng, P.; Chu, E.; Jiao, Q. Effect of Grain Size and Micromorphology of Cu Foil on the Velocity of Flyer of Exploding Foil Detonator. *Appl. Sci.* **2021**, *11*, 6598. [CrossRef]
16. Baginski, T.A.; Thomas, K.A. A Robust One-Shot Switch for High-Power Pulse Applications [EB/OL]. *IEEE Trans. Power Electron.* **2009**, *24*, 253–259. [CrossRef]
17. Baginski, T.A.; Dean, R.N.; Wild, E.J. Micromachined planar triggered spark gap switch. *IEEE Trans. Compon. Packag. Manuf. Technol.* **2011**, *1*, 1480–1485. [CrossRef]
18. Zhou, M.; Meng, Q.Y.; Han, K.H.; Qian, Y.; Qin, G.S. Design and research on exploding plane switch. *Initiat. Pyrotech.* **2012**, 12–14. Available online: https://en.cnki.com.cn/Article_en/CJFDTotal-HGPI201206005.htm (accessed on 30 December 2012).
19. Wang, R.; Cheng, T.; Zhang, Y. Research on Voltage Resistance Performance of Micro-sized Exploding Foil Plane Discharge Switch. *Initiat. Pyrotech.* **2018**, 1–4. Available online: https://caod.oriprobe.com/issues/1836321/toc.htm (accessed on 30 August 2018).

20. Li, J.; Jiao, Q.; Chu, E.; Chen, J.; Ren, W.; Li, K.; Tong, H.; Yin, G.; Wang, Y.; Zhou, M. Design, fabrication, and characterization of the modular integrated exploding foil initiator system based on low temperature co-fired ceramic technology. *Sens. Actuators A Phys.* **2020**, *315*, 112365. [CrossRef]
21. Xu, C.; Zhu, P.; Zhang, W.; Shen, R.; Ye, Y. A plasma switch induced by electro-explosion of p-n junction for mini exploding foil initiator. *IEEE Trans. Plasma Sci.* **2019**, *47*, 2710–2716. [CrossRef]
22. Xu, C.; Zhu, P.; Wang, K.; Qin, X.; Zhang, Q.; Yang, Z.; Shen, R. An electro-explosively actuated mini-flyer launcher. *Sens. Actuators A Phys.* **2019**, *292*, 17–23. [CrossRef]
23. Xu, C.; Zhu, P.; Zhang, Q.; Yang, Z.; Wang, K.; Shen, R. A shock-induced pulse power switch utilizing electro-explosion of exploding bridge wire. *IEEE Trans. Power Electron.* **2020**, *35*, 10770–10777. [CrossRef]
24. Peitian, C.; Aici, Q. Review on gas switches developed for fast linear transformer driver. *High Power Laser Part. Beams* **2012**, *24*, 1263–1268. [CrossRef]
25. Jiang, X.; Cong, P.; Sun, F.; Sun, T.; Liang, T.; Wei, H.; Wang, Z. Breakdown characteristics of a multi-gap gas switch with corona discharge for voltage balance. *High Power Laser Part. Beams* **2016**, *28*, 075009.
26. Chen, Q.; Li, Y.; Ma, T. Characterization of the super-short shock pulse generated by an exploding foil initiator. *Sens. Actuators A Phys.* **2019**, *286*, 91–97. [CrossRef]
27. Hu, B.; Jiao, J.; Zhu, P.; Wu, L.; Ye, Y.; Shen, R. Characterization of electrical explosion of schottky diode for one-shot switch applications. *Eur. Phys. J.-Appl. Phys.* **2014**, *68*, 30801. [CrossRef]
28. Liu, C.; Yang, Z.; Ding, G.; Zhou, Z.; Liu, Q.; Huang, Y.; Zheng, Y. Design, simulation, and Characterization of a low-cost in-plane spark gap microswich with dual-trigger electrode for pulsed power applications. *IEEE Trans Ind. Electrtonics* **2013**, *60*, 3240–3247. [CrossRef]
29. Baginski, M.E.; Shaffer, E.C.; Thomas, K.A.; McGuirk, J.S. Acomparison of the electrodynamics of metals under the action of large electric currents. *Int. J. Appl. Electromugn. Mech.* **2000**, *11*, 77–91. [CrossRef]
30. Lovri, M.; Schloz, F. A model for the propagation of a redox reaction through microcrystals. *Solid State Elect. Through Microcryst.* **1997**, *1*, 108–113. [CrossRef]

Review

Quantitative Assessment Methods of Early Enamel Caries with Optical Coherence Tomography: A Review

Boya Shi [1,2,*], Jiaxin Niu [1,2], Xinyue Zhou [1,2] and Xiaoyang Dong [1,2]

[1]   Department of Electronical and Information Engineering, Tiangong University, Tianjin 300387, China
[2]   Tianjin Optoelectronic Detection Technology and System Laboratory, Tianjin 300387, China
*   Correspondence: shiboya@tiangong.edu.cn

Abstract: Early detection of caries is an urgent problem in the dental clinic. Current caries detection methods do not detect early enamel caries accurately, and do not show microstructural changes in the teeth. Optical coherence tomography (OCT) can provide imaging of tiny, demineralized regions of teeth in real time and noninvasively detect dynamic changes in lesions with high resolution and high sensitivity. Over the last 20 years, researchers have investigated different methods for quantitative assessment of early caries using OCT. This review provides an overview of the principles of enamel caries detection with OCT, the methods of characterizing caries lesion severity, and correlations between OCT results and measurements from multiple histological detection techniques. Studies have shown the feasibility of OCT in quantitative assessment of early enamel lesions but they vary widely in approaches. Only integrated reflectivity and refractive index measured by OCT have proven to have strong correlations with mineral loss calculated by digital microradiography or transverse microradiography. OCT has great potential to be a standard inspection method for enamel lesions, but a consensus on quantitative methods and indicators is an important prerequisite. Our review provides a basis for future discussions.

Keywords: optical coherence tomography; early enamel caries; quantitative assessment; quantitative indicators

Citation: Shi, B.; Niu, J.; Zhou, X.; Dong, X. Quantitative Assessment Methods of Early Enamel Caries with Optical Coherence Tomography: A Review. *Appl. Sci.* **2022**, *12*, 8780. https://doi.org/10.3390/app12178780

Academic Editor: Nunzio Cennamo

Received: 28 July 2022
Accepted: 30 August 2022
Published: 31 August 2022

Publisher's Note: MDPI stays neutral with regard to jurisdictional claims in published maps and institutional affiliations.

1. Introduction

Dental caries is one of the most common chronic diseases in people worldwide because of its high incidence and the wide range of the affected population [1]. Caries is a disease of dental hard tissue caused by oral microorganisms. Normally the teeth are in a dynamic balance of alternating demineralization and remineralization, in which demineralization is the dissolution of dental minerals in the presence of acid, from which inorganic ions such as calcium and phosphate are removed, and remineralization refers to the reprecipitation and crystallization of minerals in tissues that have been demineralized [2]. If demineralization continues, caries will occur. Early enamel caries involves the demineralization of the tooth surface, and does not show substantial defects. There is no significant difference between sound enamel and early enamel caries by visual observation. At this time, it is possible to promote the repair of carious tissues through non-destructive treatment of remineralization and interrupt caries progression. However, further irreversible dental hard tissue loss will occur without interventional treatment, and then only traumatic treatment can be performed [3]. Therefore, the detection of early caries is of great importance.

Common clinical diagnostic methods mainly include visual inspection and X-ray radiography. Visual inspection lacks objectivity and has limited accuracy, while X-ray radiography can only detect severe caries lesions [4,5]. These conventional methods cannot detect early caries, which makes the early diagnosis of caries difficult and delays the best time for treatment [6]. In recent years, optical-based caries detection methods, such as Raman spectroscopy [7,8], quantitative light-induced fluorescence (QLF) [9,10] and fiber-optic transillumination [11,12], have overcome some disadvantages of traditional methods, but they do not show the internal microstructural features of teeth and cannot quantify dynamic changes in early stages of enamel lesions [13].

Optical Coherence Tomography (OCT) is a noninvasive imaging method with high resolution and high sensitivity, regarded as an "optical biopsy" [14,15]. It has a wide range of medical applications, including ophthalmology [16,17], dentistry [18,19], dermatology [20,21], and gastroenterology [22,23]. For dental caries, OCT can detect tiny, demineralized areas on the tooth surface and inside, overcoming the disadvantages of other optical detection methods [24]. In dentistry, much research has demonstrated that OCT can image dental tissue clearly and be applied to the quantitative assessment of early enamel caries [25–28]. Mineral loss is the gold standard in cariology research that determines the degree of demineralization of the lesion. Only when the quantitative index of enamel caries obtained by OCT has a definite correlation with the enamel mineral loss, the severity of early enamel caries can be quantified more objectively and accurately.

To advance the application of OCT for early caries detection, a standardized and validated method to achieve quantitative detection of early caries is required. The purpose of this paper is to present an overview of the principles, methods and applications for the quantitative evaluation of early enamel caries by OCT and to provide a basis for the establishment of a uniform clinical standard in the future. This paper first discusses the optical properties of teeth and their changes with tooth demineralization, secondly introduces methods of artificial caries preparation, and then summarizes current research advances in evaluation of early enamel lesions using different quantitative indicators and discusses the limitations of the methods.

2. Optical Properties of Teeth

Biological tissues' optical properties have important theoretical significance and application value for the diagnosis of diseases. The variation in optical properties of tooth tissue during demineralization are the basis of OCT caries detection [29].

(1) Refractive Index

Refractive index describes the reflection and refraction of light as well as the temporal properties of light transmission within a tissue. The refractive index of tooth enamel is 1.62 to 1.65 [30,31]. Demineralization in caries lesions alters the refractive index of enamel [32], and can be considered as an index of its scattering properties [33].

Since hydroxyapatite in tooth enamel is an anisotropic crystal, tooth enamel has birefringent properties [18]. When polarized light propagates through biological tissues, its polarization state is altered by the scattering properties of the tissue. By detecting changes in the polarization state of backscattered light from dental tissue, microstructural information of the tissue can be obtained, and caries lesions can be detected.

(2) Scattering Properties

Dependent on the structure of hard dental tissue, two types of scattering occur when light is transmitted inside the tooth. Dental enamel is an ordered array of hydroxyapatite crystals surrounded by a protein/lipid/water matrix. The diameter of hydroxyapatite crystals is about $30 \sim 40\,\mathrm{nm}$, and the crystals are clustered together into enamel rods with a diameter of $4\,\mu\mathrm{m}$ [34]. In the near-IR (NIR), the crystal diameter is much smaller than the source wavelength, so it is in accordance with the Rayleigh scattering law. The diameter of the enamel rods is comparable to the source wavelength, so it obeys the Mie scattering law.

Enamel has the greatest transparency in the NIR close to 1310 nm. The attenuation coefficients of sound enamel at 1310 nm and 1550 nm are 3.1 cm^{-1} and 3.8 cm^{-1} [35], respectively, which are much lower than in the visible region. Thus, the NIR spectrum is suited for tooth imaging to identify lesions [36]. Micropores form in the lesion as the mineral crystals are partially dissolved during the demineralization process, and behave as scattering centers. During initial lesion development, the scattering coefficient has an exponential increase with increasing mineral loss. As the severity of the lesion increases, the scattering gradually increases [29].

(3)   Absorption Properties

The absorption of enamel is quite weak in NIR. The absorption coefficient (μ_a) of enamel at 1310 nm is around 0.12 cm^{-1} [36], which is much smaller than the scattering coefficient. Therefore, scattering plays a major role in light transmission through the tooth.

3. Preparation of Artificial Carious Lesions

The study of early caries using OCT requires tracking of the caries process, but natural early caries takes 6–18 months, and there are many influencing factors. Therefore, the preparation of artificial caries using isolated teeth is the main way to study early caries. The key to preparing artificial caries is to simulate the chemical and microbial environment of natural caries formation. There are two main methods to prepare artificial caries: chemical and biofilm methods. Chemical methods can be used for enamel surface demineralization by adjusting the chemical state and pH of a using a partially saturated acidic buffer [37,38] or an acid gel [39]. This method is easy and the most widely used, while a biofilm method is performed by placing a cariogenic suspension on the surface of the tooth, altering the metabolic environment of the bacteria and creating a carious sample that more closely resembles natural caries [40,41]. However, the biofilm method is generally used for studies related to the pathogenesis of dental caries because it is difficult to control the experimental conditions.

4. Quantitative Assessment Methods of Early Enamel Caries with OCT

Most of the current caries detection studies based on OCT use frequency domain OCT, which eliminates axial scanning of the reference arm and overcomes the disadvantage of slow scanning speed of time-domain OCT (TD-OCT) [42]. Moreover, some functional OCT systems are employed for caries detection due to the birefringent properties of dental hard tissues. Polarization-sensitive OCT (PS-OCT) as a type of functional OCT system can provides additional polarization-sensitive information about the sample compared to conventional OCT systems [43–46]. In addition, as a new type of PS-OCT, Cross Polarization OCT (CP-OCT) not only reflects the polarization-sensitive structural information of the sample, but also better reveals the superficial microstructure of the sample by attenuating the effect of strong reflected light on the sample surface [47].

4.1. Quantitative Assessment Based on Lesion Depths

During the demineralization process, the OCT image contrast of caries is formed as the mineral content (MC) of the enamel decreases and the optical properties change [48]. Therefore, lesion depths determined by OCT images can be used to quantify early enamel caries. First, the filters are applied to remove speckle noise and enhance contrast for the OCT B-scan image, then the A-scan in the region of interest (ROI) is searched for the first pixel that exceeds the intensity threshold point, and the distance from the enamel surface to this pixel is taken as the lesion depth. The mean lesion depth is obtained by calculating the lesion depth for each A-scan in the ROI. It has been suggested to select e^{-2} times the peak intensity as the signal intensity threshold. The lesion depth measured by OCT is proportional to the demineralization time [49–53]. Le et al. [49] investigated bovine enamel caries lesions with 1310-nm PS-OCT. Figure 1 shows the PS-OCT B-scan of one of the bovine enamel blocks in the perpendicular axis. It can be seen that the lesion depth increases with the enhancement of the back-scattered intensity around the enamel surface, and the lesion severity increases. The results show the mean lesion depths varies significantly from 10 to 75 μm for demineralization over 0–4 days. Moreover, there is strong correlation (r = 0.85) between the mean depth of lesion measured by PS-OCT and polarized light microscopy (PLM). Jones et al. [50] prepared artificial caries on the occlusal surfaces by using a 14-day pH cycle model and detected them with PS-OCT and digital microradiography (DM). It was found that the image contrast between the sound and lesion area in the perpendicular axis was stronger than that in the parallel axis. The mean lesion depth of caries lesions calculated from the perpendicular axis images of the ten teeth was 150 ± 30 μm, which was highly correlated with the lesion depths obtained from DM (r = 0.811). Meng [51] and Yao et al. [52] observed an approximately linear relationship between lesion depth and demineralization time using TD-OCT. Park et al. [53] established assessment criteria for OCT using lesion depths, and conducted a concordance study between OCT and light microscopy using ex vivo teeth, which showed moderate concordance (k = 0.54, $p \leq 0.001$) with no significant difference ($p = 0.25$). Then, smooth surface in vitro and in vivo evaluations were performed using OCT and the International Caries Detection and Assessment System (ICDAS). The extent of caries was seen to vary considerably within each ICDAS category using OCT, which could effectively complement the visual assessment with ICDAS. Yavuz et al. [54] utilized 840-nm SD-OCT to assess the remineralization of artificial enamel caries. The results showed there was a significant reduction in the lesion depth after remineralization, 311.80 (344.38), 320.10 (244.36) and 312.70 (203.80) μm for the three remineralization agents, respectively. The measured lesion depth was also compared with a surface microhardness analysis but there was no correlation between the two.

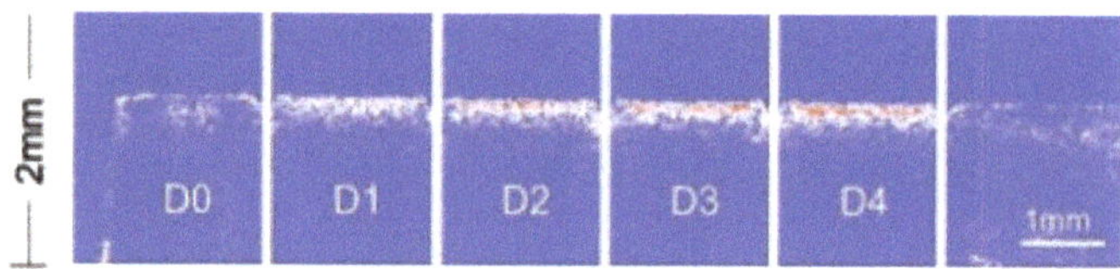

Figure 1. PS-OCT B-scan of bovine enamel in the perpendicular axis. D0 represents the sound area and D1–D4 represent areas demineralized for 1–4 days. A red-white-blue color chart was used, with red indicating strong reflectivity and blue indicating low reflectivity. This figure was adapted from [49].

Lesion depth is a common quantitative index used in early caries studies. The main challenge for its calculation is the difficulty in selecting an intensity threshold as the end point of a lesion, as the range of OCT images is relatively high. The method of selecting e^{-2} times the peak intensity as the signal intensity threshold does not always work effectively [55]. Le et al. [49] used edge-finding algorithms based on this method to measure lesion depth. Other studies designed algorithms to determine the lesion boundary in the image and obtained lesion depth [51,56,57]. A summary of the above research results is shown in Table 1.

Table 1. Results of quantifying enamel caries with lesion depths.

Reference	OCT System	Sample Preparation	Lesion Depth Result
Le et al. [49]	1310-nm PS-OCT, axial/lateral resolution 22/50 μm.	Sample number: 10 bovine enamel blocks for in vitro detection Solution: pH 4.8 demineralization solution Method: 5 windows for each block, demineralization for 0–4 days, respectively.	10–75 μm
Jones et al. [50]	1310-nm PS-OCT, axial/lateral resolution 20/30 μm.	Sample number: 15 human posterior teeth for in vitro detection Solution: pH 4.3 demineralization solution and pH 7.0 remineralization solution	150 ± 30 μm

Table 1. *Cont.*

Reference	OCT System	Sample Preparation	Lesion Depth Result
		Method: occlusal areas were exposed to 14-day pH cycling, which consisted of 6 h demineralization and 17 h remineralization a day.	
Meng et al. [51]	1310-nm TD-OCT, axial resolution 10 μm.	Sample number: 6 human isolated teeth for in vitro detection Method: demineralization for 12, 24, 48, 72, 96 and 120 h.	$d = 1.0804t + 8.6128$ d is lesion depth, t is demineralization time.
Yao et al. [52]	1310-nm TD-OCT, axial resolution 10 μm.	Sample number: 7 human permanent teeth for in vitro detection Solution: PH 4.5 demineralization solution Method: demineralization for 12, 24, 48, 72, 96, 120 h.	$d = 1.1266t + 6.7342$ d is lesion depth, t is demineralization time.
Park et al. [53]	1325-nm SS-OCT, axial/lateral resolution < 12/25 μm, imaging speed 16 kHz. 1310-nm SD-OCT, axial/lateral resolution < 7.5/15 μm, imaging speed 48–91 kHz, handheld scanning probe (Non-intraoral probe).	13 human molars, 25 human incisors, 11 premolars and 15 molars for ex vivo detection, 18 patients for in vivo detection.	Vary considerably within each ICDAS category using OCT.
Yavuz et al. [54]	840-nm SD-OCT, axial resolution 5 μm.	Sample number: 40 human enamel blocks Solution: PH 4.8 demineralization solution and PH 7.0 remineralization solution Method: 3 windows for each block, with one window demineralized for 3 days and one window remineralized for 6 days by PH cycling.	Reduction in lesion depth after remineralization for three remineralization agents: 311.80 (344.38) μm, 320.10 (244.36) μm, 312.70 (203.80) μm
Can et al. [55]	1310-nm PS-OCT, axial/lateral resolution 20/20 μm.	Sample number: 15 bovine enamel blocks for in vitro detection Solution: pH 4.9 demineralization solution Method: 3 windows for each block, demineralization for 9 days	Approximately 100 μm

Table 1. *Cont.*

Reference	OCT System	Sample Preparation	Lesion Depth Result
Jones et al. [57]	1310-nm PS-OCT, axial/lateral resolution 11/30 μm.	Sample number: 20 human posterior teeth for in vitro detection Solution: pH 4.9 demineralization solution and pH 7.0 remineralization solution Method: demineralization for 9 days and remineralization for 20 days.	Demineralization : 115 ± 16 μm, Remineralization : 104 ± 12 μm.

Lesion depths of caries can visually indicate the severity of caries to some extent. However, they can only reflect part of the characteristics of the initial stage of enamel lesions, as the amount of mineral loss may be different in a certain depth range [58]. Furthermore, there is a lack of solid criteria for determining the cut-off point to define lesion depth.

4.2. Quantitative Assessment Based on Reflectivity

Many studies have used reflectivity for the quantitative assessment of early caries, since the reflectivity of caries lesions can be obtained directly from OCT signals. A commonly used quantitative index related to reflectivity is the integrated reflectivity for caries detection.

A line profile for each lesion depth is taken from B-scans, and the integrated reflectivity (ΔR, dB × μm) can be calculated by integrating the reflectivity from the enamel surface to various depths [50]. The observed optical depth should be divided by the enamel refraction index (n = 1.63) to determine the real lesion depth when using the line profile.

Most studies have confirmed that integrated reflectivity increases after demineralization and decreases after remineralization [49,50,57,59,60]. Le et al. [49] utilized a fixed depth algorithm and an edge detection algorithm to calculate integrated reflectivity. In the first algorithm the integration was performed to a fixed depth that needed to be greater than the maximum lesion depth, while the second algorithm could obtain the depth of the lesion. The results showed both algorithms were able to detect the difference in demineralization from 0 to 4 days, except that the fixed depth algorithm yielded a higher integrated reflectivity. Jones et al. [50] calculated the mean integrated reflectivity of artificial caries prepared by applying a 14-day pH cycle model based on the perpendicular axis PS-OCT images, and the result was 450 ± 110 arbitrary units. Meanwhile, it was demonstrated that the integrated reflectivity calculated by PS-OCT was linearly correlated with the relative mineral loss determined by DM (r = 0.755). Nee et al. [59] detected demineralization around adhesive-bound orthodontic brackets in vivo using CP-OCT for a period of 1 year and acquired 2D projection images of ΔR with automated algorithms. The results indicated ΔR for both adhesives increased remarkably with time, varying in the range from 10.2 (10.5) to 29.7 (9.4) dB × μm. PS-OCT was applied to monitor the process of remineralization of caries using an acid remineralization model by Kang et al. [60]. There were significant alterations in the integrated reflectivity of the lesion region after remineralization, from 257 ± 60.2 to 168 ± 58.5 dB × μm.

Amaechi et al. [61] scanned demineralized bovine teeth using 850-nm OCT and demonstrated that ΔR of the enamel reduced with the time of demineralization. They proposed the percentage reflectivity loss ($R_\%$) as a quantitative index as follows:

$$R_\%(\mathrm{dB \times mm}) = \frac{(R_\mathrm{sound} - R_{demineralized})}{R_{sound}} \times 100\% \tag{1}$$

where R_{sound} is the reflectivity of sound enamel and $R_{demineralized}$ is the reflectivity of enamel lesion.

The results showed that $R_\%$ increased from 54.0 ± 11.27 to 86.64 ± 7.57 dB $\times$ μm with demineralization time. In a follow-on study it was demonstrated that $R_\%$ was linearly correlated with both the mineral loss determined by transverse microradiography (TMR) ($r = 1.00$) [62] and the percentage of fluorescence loss calculated (ΔQ) by QLF ($R_\% = 45.56 + \Delta Q, r = 0.963$) [63]. However, $R_\%$ is rarely applied. A summary of the above research results is shown in Table 2.

Table 2. Results of quantifyin0p-[g enamel caries with reflectivity.

Reference	OCT System	Sample Preparation	Reflectivity Result (dB$\times$μm)
Le et al. [49]	1310-nm PS-OCT, axial/lateral.	Sample number: 10 bovine enamel blocks for in vitro detection Solution: pH 4.8 demineralization solution Method: 5 windows for each block, demineralization for 0–4 days, respectively.	ΔR for the fixed depth algorithm: 100–770 ΔR for the edge detection algorithm: 20–500
Jones et al. [50]	1310-nm PS-OCT, axial/lateral resolution 20/30 μm.	Sample number: 15 human posterior teeth for in vitro detection Solution: pH 4.3 demineralization solution and pH 7.0 remineralization solution Method: occlusal areas were exposed to 14-day PH cycling, which consisted of 6 h demineralization and 17 h remineralization a day.	ΔR: 450 ± 110
Nee et al. [59]	1321-nm CP-OCT, axial/lateral resolution 11.4/80 μm, body/imaging tip of the handpiece 7×18 cm/1.5×4 cm, $6 \times 6 \times 7$ mm volume imaging time 3 s.	Sample number: Two teeth in 20 patients for in vivo detection Solution: two types of adhesives Method: The adhesives were used to bond the brackets. Demineralization around orthodontic brackets was detected every 3 months for one year.	ΔR: 10.2 (10.5)–29.7 (9.4)
Kang et al. [60]	1317-nm PS-OCT, axial/lateral resolution 20/20 μm.	Sample number: 10 bovine enamel blocks for in vitro detection Solution: pH 4.6 demineralization solution and pH 4.8 remineralization solution Method: 6 windows for each block, demineralization for 8 days and remineralization for three 4-day periods	ΔR for 8 days demineralization: 257 ± 60.2 ΔR for 4, 8, 12 days remineralization: 236 ± 73.8, 206 ± 96.0, 168 ± 58.5

Table 2. *Cont.*

Reference	OCT System	Sample Preparation	Reflectivity Result (dB×µm)
Amaechi et al. [61]	850-nm TD-OCT.	Sample number: 15 bovine teeth for in vitro detection Solution: pH 4.5 demineralization solution Method: demineralization for 3 days	$R_\%$ for 1 to 3 days: $54.0 \pm 11.27, 71.87 \pm 4.79,$ 86.64 ± 7.57
Amaechi et al. [63]	850-nm TD-OCT, axial/lateral resolution 16/10 µm.	Sample number: 15 bovine incisor teeth for in vitro detection Solution: pH 4.5 demineralization solution Method: 3 windows for each teeth. Demineralization for 24, 48, 72 h, respectively.	Reflectivity of demineralization for 0, 24, 48, 72 h: $31.86 \pm 9.30,$ $14.45 \pm 5.47, 8.66 \pm 1.96,$ 4.32 ± 2.72

In addition, other researchers have used the mean relative reflectivity (mRR) proposed in retinal OCT imaging to assess fissure caries by using 1325-nm SS-OCT [64]. The mRR is calculated from the difference between the fissure area signal and sound enamel signal. Although the mRR of demineralized fissures were at least 6 times higher than those of sound fissures at 250, 500 and 1000 µm depths beneath the surface, the mRR was unable to accurately describe lesion mineral density ($r_s = -0.31$).

In summary, the integrated reflectivity of enamel lesion calculated by OCT is linearly correlated with mineral loss. Most results confirm that the integrated reflectivity of enamel increases after demineralization, but individual studies show the opposite. Researchers often use the integrated reflectivity in combination with lesion depth to evaluate the lesion severity. However, the calculation of the integrated reflectivity suffers from similar problems to the lesion depth calculation.

4.3. Quantitative Assessment Based on Attenuation Coefficient

The attenuation coefficient is the sum of the absorption coefficient and scattering coefficient [35]. The attenuation characteristics of sound and carious enamel are different because of the alteration in optical properties of the teeth after demineralization. Hence, the attenuation coefficient can be used to quantify early enamel caries. Attenuation coefficients can be obtained by fitting the normalized A-scan signal with the Beer-Lambert law equation [65] as follows:

$$I(z) \propto \exp(-2\mu_t z) \tag{2}$$

where I is the OCT signal intensity, μ_t is the attenuation coefficient, and z is the depth beneath the tooth surface.

Mandurah et al. [66] reported that attenuation coefficients for sound areas of the samples were the smallest, ranging from 0.08 to 0.29 mm^{-1}, increasing to a range of 1.34 to 3.4 mm^{-1} after demineralization, and decreasing after remineralization with mean values of 0.81 and 0.85 mm^{-1}. Moreover, there was a strong linear regression (r = −0.97) between the μ_t measured by OCT and integrated nanohardness (INH) measured by a nanoindentation device. Hardness has been recognized as a measure of hard tissue's mineral density for a long time [67]. Maia et al. [68] studied morphological alterations between sound enamel and artificial white spot lesions in human teeth using OCT and QLF. The attenuation coefficient increases of enamel lesions ranged between 27.8% and 62.5%, while fluorescence intensity reduction ranged between 11.9% and 34.2%. Therefore, it was demonstrated that μ_t determined by OCT was more sensitive to alterations than fluorescence measured by QLF. Cara et al. [69] verified that the attenuation coefficient could be employed for the initial lesion to effectively discriminate between sound and demineralized enamel with 0.93 sensitivity and 0.96 specificity.

A weaker attenuation of the OCT signal in enamel lesions was observed in Popescu et al.'s work [70,71]. The mean attenuation coefficient was 1.35 mm^{-1} for sound enamel and 0.77 mm^{-1} for caries lesions [70]. They attributed the results to the high porosity of demineralized enamel. One possible reason for the contradictory results of the studies mentioned above is the use of different wavelengths. Mandurah et al.'s study used the 1310-nm SS-OCT system, while Popescu et al.'s study used the 850-nm OCT system. The optical properties of enamel at the two wavelength ranges are different. A summary of the above research results is shown in Table 3.

Table 3. Results of quantifying enamel caries with the attenuation coefficient.

Reference	OCT System	Sample Preparation	Attenuation Coefficient Result
Mandurah et al. [66]	1310-nm SS-OCT, axial/lateral resolution 11/17 μm, sweep rate 20 kHz.	Sample number: 24 bovine enamel blocks for in vitro detection Solution: pH 4.6 demineralization solution and two types of remineralization solution Method: 3 windows for each block, demineralization for 14 days and remineralization for 14 days.	Sound area : 0.08 – 0.29 mm^{-1}, demineralization area : 1.34–3.4 mm^{-1}, remineralization area: 0.81 and 0.85 mm^{-1} (mean).
Maia et al. [68]	930-nm SD-OCT, axial/lateral resolution 4/6 μm.	Sample number: 5 human premolar teeth for in vitro detection Solution: pH 5.0 demineralization solution Method: demineralization for 9 days.	Attenuation coefficient increase: 27.8–62.5%
Cara et al. [69]	930-nm SD-OCT, axial/lateral resolution 4/6 μm.	Sample number: 40 third molars for in vitro detection Solution: pH 4.3 demineralization solution and pH 7.4 remineralization solution Method: four groups performed 0–21 days pH cycling, consisting of 3 h demineralization and 20 h remineralization a day.	0.93 sensitivity, 0.96 specificity
Popescu et al. [70]	850-nm TD-OCT, 890 axial/lateral resolution 15/10 μm.	21 human molars and premolars for in vitro detection	Sound enamel : 0.70–2.14 mm^{-1}, caries lesion : 0.47–1.88 mm^{-1}

The Beer-Lambert equation used in the methods is based on a single scattering model. The single scattering model only considers single scattering, while the demineralization of caries enhances the effect of multiple scattering. This leads to bias of the obtained attenuation coefficients. In addition, reliably extracting attenuation coefficients from OCT signals can be affected by noise. These features diminish the utility of employing the attenuation coefficient as a marker for early enamel caries detection.

4.4. Quantitative Assessment Based on Degree of Polarization

Demineralized enamel results in rapid depolarization of polarized light in the NIR due to increased scattering [72], which has been confirmed by PS-OCT measurements [73]. Polarization imaging can provide higher contrast images of early enamel caries. Since OCT is an interferometric imaging method, only the contribution of fully polarized light can be measured. Thus, the degree of polarization within a single speckle is always equal to 1. However, when depolarized, the polarization state of the adjacent speckles is uncorrelated. Therefore, the degree of polarization uniformity (DOPU) has been proposed to assess carious lesions [74]. DOPU can be derived by an averaging of Stokes vectors over adjacent speckles, as follows:

$$\text{DOPU} = \sqrt{Q_{mean}^2 + U_{mean}^2 + V_{mean}^2} \tag{3}$$

where Q_{mean}, U_{mean} and V_{mean} are the mean values of the Stokes vector elements within a certain evaluation kernel. It can be seen that the value of DOPU depends on the number of speckles in the chosen kernel.

The combination of the DOPU algorithm and PS-OCT was first applied to detect carious lesions by Golde et al. [74], and has been used for ophthalmologic research [75]. They measured three tooth samples with different proximal lesions, and the significant DOPU contrast provided better identification of lesions in comparison with reflectivity images. Furthermore, the effect of different DOPU evaluation kernel sizes on the resulting contrast was investigated. In a following study, they improved the DOPU algorithm by noise-immune processing, and adopted this approach to examine two tooth samples with stains and occlusal lesions [76]. Then, the research group measured the DOPU of bovine enamel at different stages of demineralization by using PS-OCT, and compared it with lesion depth obtained from PLM measurements [77]. The results showed that there was no depolarization in sound enamel, but an increased depolarization after 15 days of demineralization, corresponding to a decrease in DOPU. There was a high linear correlation ($R^2 = 0.7118$) between the DOPU and measured lesion depth with PLM, as shown in Figure 2. The summary of the above research results is shown in Table 4.

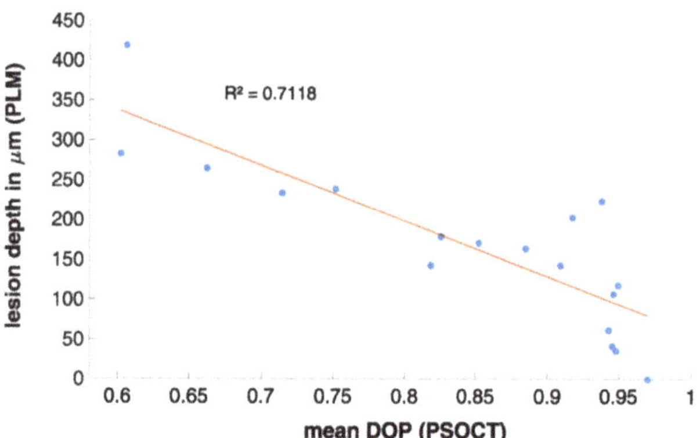

Figure 2. Correlation between the calculated mean DOP by PS-OCT and determined lesion depths by PLM. This figure was adapted from [77].

While the above results indicate the feasibility of assessing the demineralization stage by DOPU, there is a need to investigate the validity of DOPU at various polarization changes and the correlation between DOPU and mineral loss measured by TMR for further studies.

Table 4. Results of quantifying enamel caries with degree of polarization.

Reference	OCT System	Sample Preparation	DOPU Result
Golde et al. [74]	1310-nm SD PS-OCT, axial/lateral resolution 15.1/15.6 µm, sweep rate 50 kHz.	3 human molar teeth with different proximal lesions	The significant DOPU contrast provided better identification of lesions in comparison with the reflectivity images.
Tetschke et al. [77]	1310-nm SD PS-OCT, axial/lateral resolution 15.1/15.6 µm, sweep rate 50 kHz.	Sample number: 18 bovine enamel blocks for in vitro detection Solution: pH 4.95 demineralization solution Method: demineralization for 49 days.	0.6–0.97

4.5. Quantitative Assessment Based on Refractive Index

Demineralization causes a change in the refractive index of enamel, and accurate measurement of this change can assist in the identification of early caries [32]. The refractive index of teeth can be determined with the optical path-length matching method using OCT [78]. Samples are put onto a metal plate to acquire OCT images. The depth position of the reflection surface of the metal plate before the sample is placed is Z_0. After adding samples, the depth positions of the sample surface and the metal plate surface are Z_1 and Z'_0, respectively. The thickness of the sample is $Z_1 - Z_0$. Then, the refractive index of teeth is determined by [79]:

$$n = \frac{Z_1 - Z'_0}{Z_1 - Z_0} \qquad (4)$$

Hariri et al. [80] measured refractive index of sound bovine enamel, demineralized for 2 months, and remineralized for 2 months, by 1310-nm SS-OCT with an axial/lateral resolution 11/17 µm, and analyzed mineral content by TMR. The results showed that at an n range between 1.52 and 1.63, the mineral content ranged between 50 and 87 (vol.%). This indicates there were strong positive linear correlations between n and mineral content in both demineralized enamel ($R^2 = 0.89$) and remineralized enamel ($R^2 = 0.86$). However, this method required sectioning of the sample to measure the refractive index, which is destructive and cannot be applied in clinical practice.

4.6. Quantitative Assessment Based on Scattering Coefficient

Since the scattering properties of the enamel changes significantly after demineralization, the scattering coefficient can serve as an indicator of enamel lesion severity. A single scattering model combined with dynamic focusing can be used to determine the scattering coefficients of sound and carious enamel [81]. The OCT signal intensity is:

$$I(z) \propto \sqrt{\frac{e^{-2\mu_s z}}{\left[1 + \left(\frac{z - z_{cf}}{z_R}\right)^2\right]}} \qquad (5)$$

where z is the depth, z_{cf} is the focal plane position, z_R is Rayleigh length, and $I(z)$ is the OCT depth profile.

Tsai et al. [81] applied acid gel to demineralize enamel and scanned the sample in vitro before and after demineralization using 850-nm SD-OCT with an axial/lateral resolution 3/4 µm. The estimated scattering coefficient is shown in Figure 3. The scattering coefficient increased with the demineralization time and leveled out at times greater than 120 s. Moreover, the average scattering coefficients were 4.60 mm^{-1} and 8.46 mm^{-1} for sound and carious enamel, respectively.

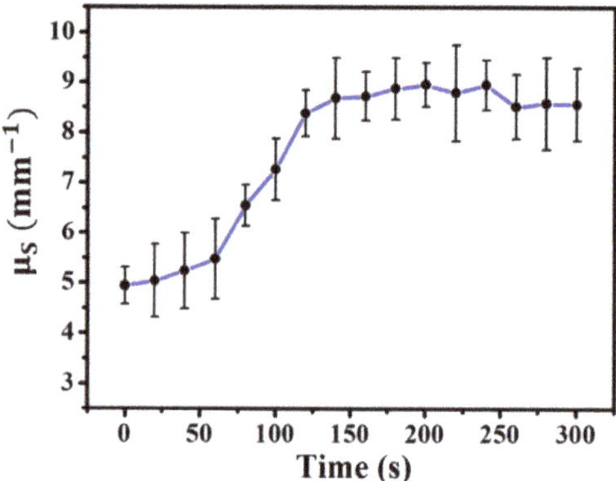

Figure 3. Variation of scattering coefficient with demineralization time. This figure was adapted from [81] and was created with Microsoft Word (Microsoft Corp., Redmond, WA, USA).

However, as mentioned above, the single scattering model used above is inaccurate for enamel caries. According to the optical properties of the caries lesion, it shows a significant growth in the scattering coefficient of enamel during the production of the initial lesion. Hence, if an accurate scattering coefficient is used for the quantitative assessment of early caries, early demineralization can be detected more sensitively. However, there are few studies using scattering coefficient to quantitatively evaluate early caries.

4.7. Quantitative Assessment Based on the Surface Roughness of Enamel

Acid or bacterial erosion alters the surface roughness of enamel. The root mean square of the surface roughness can be expressed as [81]:

$$R_q = \sqrt{\frac{1}{n}\sum(z_b - z_t)^2} \tag{6}$$

where n is the total number of A-scans, z_t and z_b are the surface and underlying depth positions of lesion area in A-scan, respectively.

Tsai et al. [81] estimated the surface roughness of the demineralized enamel, as shown in Figure 4. The surface roughness of the enamel increased gradually with demineralization time, and tended to level out, varying from 5.11 μm to 31.7 μm. Although the results demonstrated that the surface roughness could be applied for the detection of early caries, there were estimation errors compared with the results of scanning electron microscopy (SEM), and the effect of artificial caries and natural caries on enamel surface roughness may be different. In addition, there are few relevant studies.

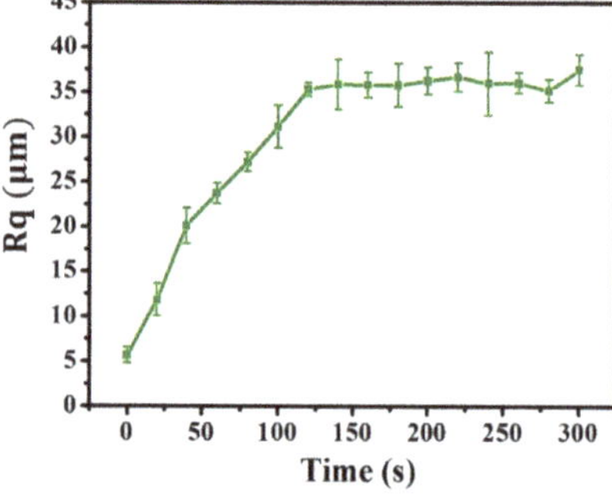

Figure 4. Variation of the surface roughness with demineralization time. This figure was adapted from [81] and was created with Microsoft Word (Microsoft Corp., Redmond, WA, USA).

4.8. Quantitative Assessment Based on the Volume of Residual Enamel

Demineralization caused by caries lesions changes the volume of residual enamel. Wijesinghe et al. [82] obtained cross-sectional images of sound, partially demineralized and completely demineralized teeth in vitro by using 1310-nm SD-OCT with an axial/lateral resolution 6/25 μm, and measured the volume of residual enamel with an automated calculation method based on pixel intensity. The volumetric evaluation algorithm is shown in Figure 5. For the precise selection of residual enamel, an image window is applied to the 2D OCT images, as shown in Figure 5a. Then, the pixels that satisfy the pre-determined intensity threshold range are selected as shown in Figure 5b. Finally, a 3D OCT volumetric image is obtained as shown in Figure 5c. The volume of residual enamel is determined by:

$$V_{tot} = \left(N_1 \times l_x \times l_y \right) \times l_z + \left(N_2 \times l_x \times l_y \right) \times l_z + \cdots + \left(N_n \times l_x \times l_y \right) \times l_z \qquad (7)$$

where $N_i (i = 1, 2, \ldots, n)$ is the number of pixels in each window that satisfy the predetermined intensity cut-off points, n is the number of 2D images contained in the 3D image. l_x, l_y and l_z are the pixel sizes in the x, y and z directions, respectively.

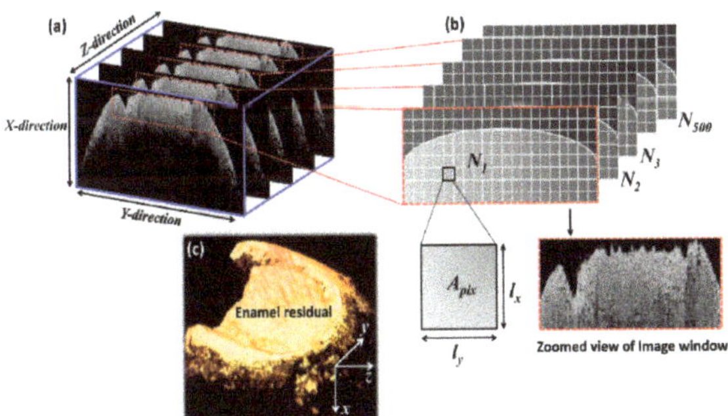

Figure 5. Evaluation algorithm for the volume of residual enamel. (**a**) 2D images with the applied image window; (**b**) Pixels that meet the predetermined intensity cut-off points; (**c**) 3D volumetric image. This figure was adapted from [82].

The volume of residual enamel for carious samples, partially demineralized samples and sound samples ranged from 12.26 to 28.72 mm^3. The progression of dental caries is determined by detecting changes in the volume of residual enamel. When reduction in tooth volume is identified, medication can be taken immediately to inhibit the development of caries. The key in this method is the determination of threshold parameters, which requires the evaluation and standardization of volumetric information for multiple in vivo teeth to enhance accuracy.

4.9. Quantitative Assessment Based on the Dehydration Parameter

Recent research has revealed the impact of hydration on OCT images by conventional polarization-insensitive OCT [83,84]. A method for assessing early enamel lesions with the dehydration parameter (DH) based on the integrated reflectivity has been presented [85]. The dehydration parameter refers to the positive difference between the two OCT signals of tooth enamel under dry and hydrated conditions, i.e., the integrated area between the two signals.

Nazari et al. [85] detected sound and demineralized bovine enamel blocks after 3, 9, and 15 days of demineralization using 1310-nm SS-OCT with axial/lateral resolution 11/17 μm, and calculated the dehydration parameters. The experimental results showed DH for sound and demineralized bovine enamel ranged from 272(204) to 3304(751), and a strong linear correlation ($R^2 = 0.9922$) between the dehydration parameter and the square root of demineralization day. The benefit of this method is that the DH calculation does not involve determination of the cut-off depth, and the evaluation results are not influenced by surface reflections. Although this method has the potential to quantitatively assess early enamel caries, there are few relevant studies, and optimization of the methodology requires evaluation of large amounts of demineralized and remineralized enamel.

5. Conclusions

In conclusion, the common aim of the discussed studies was to investigate and improve the capabilities of OCT in monitoring the pathophysiological process of early enamel caries and remineralization. Researchers have proposed multiple quantitative indicators for the assessment of early enamel caries and have studied correlations with the results of multiple histological detection techniques. By far the most widely used quantitative indicators include lesion depth, integrated reflectivity and attenuation coefficients. The differences in quantitative results are mainly attributed to the use of different systems, different methods and different sample preparation. Among them, there is a high linear correlation between depths of enamel lesions determined by OCT and DM, as well as between integrated reflectivity calculated by OCT and mineral loss determined by DM. However, the assessment methods using these three quantitative indicators still have certain limitations and lack objectivity and accuracy. It is a remarkable fact that the scattering coefficient of tooth enamel is very sensitive to changes in mineral content and increases significantly with increasing mineral loss at the initial stages of the enamel caries process. Therefore, the early detection of enamel lesions using the scattering coefficient based on OCT is a potential research direction. Meanwhile, most studies on quantitative assessment of enamel caries have been performed in vitro due to the limitations of the probe, and in vivo studies are mainly focused on the buccal and incisal/occlusal surfaces of premolars and anterior teeth. With the development of intraoral probes [4,86], the problem of device availability is being solved slowly. However, a very important issue is the lack of consensus on the method, and OCT has not been applied for the clinical diagnosis of early caries. Although OCT has made great progress as a noninvasive imaging method for quantitatively assessing early enamel caries, efforts to standardize rigorous methodology in future research are crucial for detection, diagnosis, and treatment guidance of early enamel lesions using OCT.

Author Contributions: Conceptualization, all authors; methodology, B.S., J.N., X.Z.; formal analysis, B.S., J.N.; investigation, B.S., J.N.; resources, all authors; data curation, B.S., J.N.; writing—original draft preparation, B.S., J.N.; writing—review and editing, all authors; visualization, B.S., J.N.; supervision, X.Z., X.D.; project administration, B.S., J.N.; funding acquisition, B.S., J.N. All authors have read and agreed to the published version of the manuscript.

Funding: This research was funded by the Natural Science Foundation of Tianjin (grant number 19JCYBJC16200) and the Science and Technology Program of Tianjin (grant number 20YDTPJC01530).

Institutional Review Board Statement: Not applicable.

Informed Consent Statement: Not applicable.

Data Availability Statement: All the abovementioned data can be found in the references.

Conflicts of Interest: The authors declare no conflict of interest.

References

1. Baelum, V.; Fejerskov, O. Caries diagnosis: A mental resting place on the way to intervention. In *Dental Caries–The Disease and Its Clinical Management*; Blackwell Munksgaard: London, UK, 2003; pp. 101–110.
2. Dirks, O.B. Posteruptive Changes in Dental Enamel. *J. Dent. Res.* **1966**, *45*, 503–511. [CrossRef]
3. Tenbosch, J.J.; Verdonschot, E.H.; Vaarkamp. Light propagation through teeth containing simulated caries lesions. *Phys. Med. Biol.* **1995**, *40*, 1375–1387.
4. Schneider, H.; Ahrens, M.; Strumpski, M.; Rüger, C.; Häfer, M.; Hüttmann, G.; Theisen-Kunde, D.; Schulz-Hildebrandt, H.; Haak, R. An Intraoral OCT Probe to Enhanced Detection of Approximal Carious Lesions and Assessment of Restorations. *J. Clin. Med.* **2020**, *9*, 3257. [CrossRef] [PubMed]
5. Schneider, H.; Park, K.J.; Häfer, M.; Rüger, C.; Schmalz, G.; Krause, F.; Schmidt, J.; Ziebolz, D.; Haak, R. Dental Applications of Optical Coherence Tomography (OCT) in Cariology. *Appl. Sci.* **2017**, *7*, 472. [CrossRef]

6. Brouwer, F.; Askar, H.; Paris, S.; Schwendicke, F. Detecting Secondary Caries Lesions: A Systematic Review and Meta-analysis. *J. Dent. Res.* **2016**, *95*, 143–151. [CrossRef] [PubMed]
7. Jones, R.S.; Huynh, G.D.; Jones, G.C.; Fried, D. Near-infrared transillumination at 1310-nm for the imaging of early dental decay. *Opt. Exp.* **2003**, *11*, 2259–2265. [CrossRef]
8. Schaefer, G.; Pitchika, V.; Litzenburger, L.; Hickel, R.; Kühnisch, J. Evaluation of occlusal caries detection and assessment by visual inspection, digital bitewing radiography and near-infrared light transillumination. *Clin. Oral Investig.* **2018**, *22*, 2431–2438. [CrossRef]
9. Alammari, M.R.; Smith, P.W.; De Josselin De Jong, E.; Higham, S.M. Quantitative light-induced fluorescence (QLF): A tool for early occlusal dental caries detection and supporting decision making in vivo. *J. Dent.* **2013**, *41*, 127–132. [CrossRef]
10. Gomez, J. Detection and diagnosis of the early caries lesion. *BMC Oral Health* **2015**, *15*, 3. [CrossRef]
11. Abogazalah, N.; Ando, M. Alternative methods to visual and radiographic examinations for approximal caries detection. *J. Oral Sci.* **2017**, *59*, 315–322. [CrossRef]
12. Pretty, I.A. Caries detection and diagnosis: Novel technologies. *J. Dent.* **2006**, *34*, 727–739. [CrossRef]
13. Patil, C.A.; Bosschaart, N.; Keller, M.D.; Leeuwen, T.G.; Mahadevan-Jansen, A. Combined Raman spectroscopy and optical coherence tomography device for tissue characterization. *Opt. Lett.* **2008**, *33*, 1135–1137. [CrossRef] [PubMed]
14. Qiao, W.; Chen, Z. All-optically integrated photoacoustic and optical coherence tomography: A review. *J. Innov. Opt. Health Sci.* **2017**, *10*, 1730006. [CrossRef]
15. Tearney, G.J.; Brezinski, M.E.; Bouma, B.E.; Boppart, S.A.; Pitvis, C.; Southern, J.F.; Fujimoto, J.G. In vivo endoscopic optical biopsy with optical coherence tomography. *Science* **1997**, *276*, 2037–2039. [CrossRef]
16. Hangai, M.; Ojima, Y.; Gotoh, N.; Inoue, R.; Yasuno, Y.; Makita, S.; Yamanari, M.; Yatagai, T.; Kita, M.; Yoshimura, N. Three-dimensional Imaging of Macular Holes with High-speed Optical Coherence Tomography. *Ophthalmology* **2007**, *114*, 763–773. [CrossRef] [PubMed]
17. Yasuno, Y.; Hong, Y.; Makita, S. In vivo high-contrast imaging of deep posterior eye by 1-μm swept source optical coherence tomography and scattering optical coherence angiography. *Opt. Exp.* **2007**, *15*, 6121–6139. [CrossRef]
18. Baumgartner, A.; Dichtl, S.; Hitzenberger, C.K.; Sattmann, H.; Robl, B.; Moritz, A.; Fercher, A.F.; Sperr, W. Polarization–Sensitive Optical Coherence Tomography of Dental Structures. *Caries Res.* **2000**, *14*, 59–69. [CrossRef] [PubMed]
19. Colston, B.W.; Sathyam, U.S.; DaSilva, L.B.; Everett, M.J.; Stroeve, P.; Otis, L.L. Dental OCT. *Opt. Exp.* **1998**, *3*, 230–238. [CrossRef]
20. Pagnoni, A.; Knuettel, A.; Welker, P.; Rist, M.; Stoudemayer, T.; Kolbe, L.; Sadiq, I.; Kligman, A.M. Optical coherence tomography in dermatology. *Skin Res. Technol.* **1999**, *5*, 83–87. [CrossRef]
21. Pierce, M.C.; Strasswimmer, J.; Park, H.; Cense, B.; Boer, J.F. Birefringence measurements in human skin using polarization-sensitive optical coherence tomography. *J. Biomed. Opt.* **2004**, *9*, 287–291. [CrossRef] [PubMed]
22. Evans, J.A.; Poneros, J.M.; Bouma, B.E.; Bressner, J.; Halpern, E.F.; Shishkov, M.; Lauwers, G.Y.; Kenudson, M.M.; Nishioka, N.S.; Tearney, G.J. Optical coherence tomography to identify intramucosal carcinoma and high-grade dysplasia in Barrett's esophagus. *Clin. Gastroenterol. Hepatol.* **2006**, *4*, 38–43. [CrossRef]
23. Poneros, J.M.; Brand, S.; Bouma, B.E.; Tearney, G.J.; Compton, C.C.; Nishioka, N.S. Diagnosis of Specialized Intestinal Metaplasia by Optical Coherence Tomography. *Gastroenterology* **2001**, *51*, 7–12. [CrossRef] [PubMed]
24. Yang, Z.; Shang, J.; Liu, C.; Zhang, J.; Hou, F.; Liang, Y. Intraoperative imaging of oral-maxillofacial lesions using optical coherence tomography. *J. Innov. Opt. Health Sci.* **2020**, *13*, 2050011. [CrossRef]
25. Na, J.; Baek, J.H.; Choi, E.S.; Ryu, S.Y.; Chang, J.; Lee, C.S.; Lee, B.H. Assessment of dental-caries using optical coherence tomography. *Int. Soc. Opt. Eng. SPIE* **2006**, *6137*, 18–27.
26. Na, J.; Baek, J.H.; Ryu, S.Y.; Lee, C.; Lee, B.H. Tomographic imaging of incipient dental-caries using optical coherence tomography and comparison with various modalities. *Opt. Rev.* **2009**, *16*, 426–431. [CrossRef]
27. Holtzman, J.S.; Osann, K.; Pharar, J. Ability of Optical Coherence Tomography to Detect Caries Beneath Commonly Used Dental Sealants. *Lasers Surg. Med.* **2010**, *42*, 752–759. [CrossRef]
28. Shimada, Y.; Sadr, A.; Burrow, M.F.; Tagami, J.; Ozawa, N.; Sumi, Y. Validation of swept-source optical coherence tomography (SS-OCT) for the diagnosis of occlusal caries. *J. Dent.* **2010**, *38*, 655–665. [CrossRef]
29. Darling, C.L.; Huynh, G.; Fried, D. Light scattering properties of natural and artificially demineralized dental enamel at 1310 nm. *J. Biomed. Opt.* **2006**, *11*, 034023. [CrossRef]
30. Ohmi, M.; Ohnishi, Y.; Yoden, K.; Haruna, M. In vitro simultaneous measurement of refractive index and thickness of biological tissue by the low coherence interferometry. *IEEE Trans. Biomed. Eng.* **2000**, *47*, 1266–1270. [CrossRef]
31. Wang, X.J.; Milner, T.E.; Boer, J.F.d.; Zhang, Y.; Pashley, D.H.; Nelson, J.S. Characterization of dentin and enamel by use of optical coherence tomography. *Appl. Opt.* **1999**, *38*, 2092–2096. [CrossRef]
32. Besic, F.C.; Wiemann, M.R. Dispersion Staining, Dispersion, and Refractive Indices in Early Enamel Caries. *J. Dent. Res.* **1972**, *51*, 973–985. [CrossRef]
33. Knuettel, A.R.; Bonev, S.M.; Knaak, W. New method for evaluation of in vivo scattering and refractive index properties obtained with optical coherence tomography. *J. Biomed. Opt.* **2004**, *9*, 265–273. [CrossRef] [PubMed]
34. Fried, D.; Glena, R.E.; Featherstone, J.D.B.; Seka, W. Nature of light scattering in dental enamel and dentin at visible and near-infrared wavelengths. *Appl. Opt.* **1995**, *34*, 1278–1285. [CrossRef] [PubMed]

35. Jones, R.; Fried, D. Attenuation of 1310-nm and 1550-nm laser light through sound dental enamel. *Lasers Dent. VIII* **2002**, *4610*, 187–190.

36. Spitzer, D.; Bosch, J.T. The absorption and scattering of light in bovine and human dental enamel. *Calcif. Tiss. Res.* **1975**, *17*, 129–137. [CrossRef] [PubMed]

37. Chan, K.H.; Chan, A.C.; Darling, C.L.; Fried, D. Methods for Monitoring Erosion Using Optical Coherence Tomography. *Lasers Dent. XIX* **2013**, *8566*, 35–40.

38. Lee, C.; Darling, C.L.; Fried, D. Polarization-sensitive optical coherence tomographic imaging of artificial demineralization on exposed surfaces of tooth roots. *Dent. Mater.* **2009**, *25*, 721–728. [CrossRef]

39. Şen, S.; Erber, R.; Deurer, N.; Orhan, G.; Lux, C.J.; Zingler, S. Demineralization Detection in Orthodontics Using an Ophthalmic Optical Coherence Tomography Device Equipped with a Multicolor Fluorescence Module. *Clin. Oral Investig.* **2020**, *24*, 2579–2590. [CrossRef]

40. Azevedo, C.S.; Trung, L.C.E.; Simionato, M.R.L.; Freitas, A.Z.; Matos, A.B. Evaluation of caries-affected dentin with optical coherence tomography. *Braz. Oral Res.* **2011**, *25*, 407–413. [CrossRef] [PubMed]

41. Freitas, A.Z.; Zezell, D.M.; Mayer, M.P.A.; Ribeiro, A.C.; Gomes, A.S.L.; Jr., N.D.V. Determination of dental decay rates with optical coherence tomography. *Laser Phys. Lett.* **2009**, *6*, 896–900. [CrossRef]

42. Espigares, J.; Sadr, A.; Hamba, H.; Shimada, Y.; Otsuki, M.; Tagami, J.; Sumi, Y. Assessment of natural enamel lesions with optical coherence tomography in comparison with microfocus X-ray computed tomography. *J. Med. Imag.* **2015**, *2*, 014001. [CrossRef] [PubMed]

43. Hee, M.R. Optical Coherence Tomography of the Human Retina. *Arch. Ophthalmol.* **1995**, *113*, 325–332. [CrossRef] [PubMed]

44. Kurokawa, K.; Sasaki, K.; Makita, S.; Yamanari, M.; Cense, B.; Yasuno, Y. Simultaneous high-resolution retinal imaging and high-penetration choroidal imaging by one-micrometer adaptive optics optical coherence tomography. *Opt. Exp.* **2010**, *18*, 8515–8527. [CrossRef]

45. Boer, J.F.; Milner, T.E.; Gemert, M.J.; Nelson, J.S. Two-dimensional birefringence imaging in biological tissue by polarization-sensitive optical coherence tomography. *Opt. Lett.* **1997**, *22*, 934–936. [CrossRef]

46. Saxer, C.E.; Boer, J.F.; Hyle Park, B.H.; Zhao, Y.; Chen, Z.; Nelson, J.S. High-speed fiber-based polarization-sensitive optical coherence tomography of in vivo human skin. *Opt. Lett.* **2000**, *25*, 1355–1357. [CrossRef] [PubMed]

47. Chan, K.H.; Tom, H.; Lee, R.C.; Kang, H.; Simon, J.C.; Staninec, M.; Darling, C.L.; Pelzner, R.B.; Fried, D. Clinical monitoring of smooth surface enamel lesions using CP-OCT during nonsurgical intervention. *Lasers Surg. Med.* **2016**, *48*, 915–923. [CrossRef] [PubMed]

48. Chen, Y.; Otis, L.; Piao, D.; Zhu, Q. Characterization of dentin, enamel, and carious lesions by a polarization-sensitive optical coherence tomography system. *Appl. Opt.* **2005**, *44*, 2041–2048. [CrossRef]

49. Le, M.H.; Darling, C.L.; Fried, D. Automated analysis of lesion depth and integrated reflectivity in PS-OCT scans of tooth demineralization. *Lasers Surg. Med.* **2010**, *42*, 62–68. [CrossRef] [PubMed]

50. Jones, R.S.; Darling, C.L.; Featherstone, J.D.B.; Fried, D. Imaging Artificial Caries on the Occlusal Surfaces with Polarization-Sensitive Optical Coherence Tomography. *Caries Res.* **2006**, *40*, 81–89. [CrossRef]

51. Meng, Z.; Yao, X.; Yao, H.; Liu, T.; Li, Y.; Wang, G. Detecting early artificial caries by using optical coherence tomography. *Chin. J. Lasers* **2010**, *37*, 2709–2713. [CrossRef]

52. Yao, H.; Li, Y.; Wang, G.; Yao, X.; Meng, Z.; Jin, S.; Liang, Y.; Zhang, L.; Liu, T. Quantification detecting artificial early caries with an All-fiber-OCT system. *Int. J. Biomed. Eng.* **2009**, *37*, 65–69+66.

53. Park, K.J.; Schneider, H.; Ziebolz, D.; Krause, F.; Haak, R. Optical coherence tomography to evaluate variance in the extent of carious lesions in depth. *Lasers Med. Sci.* **2018**, *33*, 1573–1579. [CrossRef]

54. Yavuz, B.S.; Kargul, B. Comparative evaluation of the spectral-domain optical coherence tomography and microhardness for remineralization of enamel caries lesions. *Dent. Mater. J.* **2021**, *40*, 1115–1121. [CrossRef] [PubMed]

55. Can, A.M.; Darling, C.L.; Ho, C.; Fried, D. Non-destructive assessment of inhibition of demineralization in dental enamel irradiated by a λ = 9.3-μm CO2 laser at ablative irradiation intensities with PS-OCT. *Lasers Surg. Med.* **2008**, *40*, 342–349. [CrossRef]

56. Staninec, M.; Douglas, S.M.; Darling, C.L.; Chan, K.; Kang, H.; Lee, R.C.; Fried, D. Non-destructive clinical assessment of occlusal caries lesions using near-IR imaging methods. *Lasers Surg. Med.* **2011**, *43*, 951–959. [CrossRef] [PubMed]

57. Jones, R.S.; Fried, D. Remineralization of Enamel Caries Can Decrease Optical Reflectivity. *J. Dent. Res.* **2006**, *85*, 804–808. [CrossRef] [PubMed]

58. Jones, R.S.; Darling, C.L.; Featherstone, J.D.B.; Fried, D. Remineralization of in vitro dental caries assessed with polarization-sensitive optical coherence tomography. *J. Biomed. Opt.* **2006**, *11*, 014016. [CrossRef] [PubMed]

59. Nee, A.; Chan, K.; Kang, H.; Staninec, M.; Darling, C.L.; Fried, D. Longitudinal monitoring of demineralization peripheral to orthodontic brackets using cross polarization optical coherence tomography. *J. Dent.* **2014**, *42*, 547–555. [CrossRef]

60. Kang, H.; Darling, C.L.; Fried, D. Nondestructive monitoring of the repair of enamel artificial lesions by an acidic remineralization model using polarization-sensitive optical coherence tomography. *Dent. Mater.* **2012**, *28*, 488–494. [CrossRef]

61. Amaechi, B.T.; Higham, S.M.; Podoleanu, A.G.; Rogers, J.A.; Jackson, D.A. Use of optical coherence tomography for assessment of dental caries: Quantitative procedure. *J. Oral Rehabil.* **2001**, *28*, 1092–1093. [CrossRef]

62. Amaechi, B.T.; Podoleanu, A.; Komarov, G.; Higham, S.M.; Jackson, D.A. Optical coherence tomography for dental caries detection and analysis. *Lasers Dent. VIII* **2002**, *4610*, 100–108.

63. Amaechi, B.T.; Podoleanu, A.G.; Higham, S.M.; Jackson, D.A. Correlation of quantitative light-induced fluorescence and optical coherence tomography applied for detection and quantification of early dental caries. *J. Biomed. Opt.* **2003**, *8*, 642–647. [CrossRef] [PubMed]

64. Liu, X.; Jones, R.S. Evaluating a novel fissure caries model using swept source optical coherence tomography. *Dent. Mater. J.* **2013**, *32*, 906–912. [CrossRef]
65. Sowa, M.G.; Popescu, D.P.; Werner, J.; Hewko, M.; Ko, A.C.T.; Payette, J.; Dong, C.C.S.; Cleghorn, B.; Choo-Smith, L.P. Precision of Raman depolarization and optical attenuation measurements of sound tooth enamel. *Anal. Bioanal. Chem.* **2007**, *387*, 1613–1619. [CrossRef] [PubMed]
66. Mandurah, M.; Sadr, A.; Shimada, Y.; Kitasako, Y.; Nakashima, S.; Bakhsh, T.A.; Tagami, J.; Sumi, Y. Monitoring remineralization of enamel subsurface lesions by optical coherence tomography. *J. Biomed. Opt.* **2013**, *18*, 046006. [CrossRef]
67. Featherstone, J.D.B.; Cate, J.M.; Shariati, M.; Arends, J. Comparison of artificial caries-like lesions by quantitative microradiography and microhardness profiles. *Caries Res.* **1983**, *17*, 385–391. [CrossRef] [PubMed]
68. Maia, A.M.A.; Freitas, A.Z.; Campello, S.; Gomes, A.S.L.; Karlsson, L. Evaluation of dental enamel caries assessment using quantitative light induced fluorescence and optical coherence tomography. *J. Biophotonics* **2016**, *9*, 596–602. [CrossRef]
69. Cara, A.C.B.; Zezell, D.M.; Ana, P.A.; Maldonado, E.P.; Freitas, A.Z. Evaluation of two quantitative analysis methods of optical coherence tomography for detection of enamel demineralization and comparison with microhardness. *Lasers Surg. Med.* **2014**, *46*, 666–671. [CrossRef]
70. Popescu, D.P.; Sowa, M.G.; Hewko, M.D.; Choo-Smith, L.P. Assessment of early demineralization in teeth using the signal attenuation in optical coherence tomography images. *J. Biomed. Opt.* **2008**, *13*, 054053. [CrossRef]
71. Sowa, M.G.; Popescu, D.P.; Friesen, J.R.; Hewko, M.D.; Choo-Smith, L.-P. A comparison of methods using optical coherence tomography to detect demineralized regions in teeth. *J. Biophotonics* **2011**, *4*, 814–823. [CrossRef]
72. Darling, C.L.; Fried, D. Polarized light propagation through sound and carious enamel at 1310-nm. *Proc. SPIE–Int. Soc. Opt. Eng.* **2006**, *6137*, 151–158.
73. Everett, M.J.; Colston, B.W.; Sathyam, U.S.; Silva, L.B.; Fried, D.; Featherstone, J.D. Noninvasive diagnosis of early caries with polarization-sensitive optical coherence tomography (PS-OCT). *Lasers Dent. V SPIE* **2022**, *3593*, 177–182.
74. Golde, J.; Tetschke, F.; Walther, J.; Rosenauer, T.; Hempel, F.; Hannig, C.; Koch, E.; Kirsten, L. Detection of carious lesions utilizing depolarization imaging by polarization sensitive optical coherence tomography. *J. Biomed. Opt.* **2018**, *23*, 071201–071208. [CrossRef] [PubMed]
75. Götzinger, E.; Pircher, M.; Geitzenauer, W.; Ahlers, C.; Baumann, B.; Michels, S.; Schmidt-Erfurth, U.; Hitzenberger, C.K. Retinal pigment epithelium segmentation by polarization sensitive optical coherence tomography. *Opt. Exp.* **2008**, *16*, 16410–16422. [CrossRef]
76. Golde, J.; Tetschke, F.; Vosahlo, R.; Walther, J.; Hannig, C.; Koch, E.; Kirsten, L. Assessment of occlusal enamel alterations utilizing depolarization imaging based on PS-OCT. *Biomed. Opt. Imag. SPIE* **2019**, *11078*, 11078_24.
77. Tetschke, F.; Golde, J.; Rosenauer, T.; Basche, S.; Walther, J.; Kirsten, L.; Koch, E.; Hannig, C. Correlation between Lesion Progression and Depolarization Assessed by Polarization-Sensitive Optical Coherence Tomography. *Appl. Sci.* **2020**, *10*, 2971. [CrossRef]
78. Meng, Z.; Yao, X.; Yao, H.; Liang, Y.; Liu, T.; Li, Y.; Wang, G. Measurement of the refractive index of human teeth by optical coherence tomography. *J. Biomed. Opt.* **2009**, *14*, 034010. [CrossRef]
79. Hariri, I.; Sadr, A.; Shimada, Y.; Nakashima, S.; Sumi, Y.; Tagami, J. Relationship between Refractive Index and Mineral Content of Enamel and Dentin Using SS-OCT and TMR. *Lasers Dent. XVIII* **2012**, *8208*, 117–122.
80. Hariri, I.; Sadr, A.; Nakashima, S.; Shimada, Y.; Tagami, J.; Sumi, Y. Estimation of the Enamel and Dentin Mineral Content from the Refractive Index. *Caries Res.* **2013**, *47*, 18–26. [CrossRef]
81. Tsai, M.T.; Wang, Y.L.; Yeh, T.W.; Lee, H.C.; Chen, W.J.; Ke, J.L.; Lee, Y.J. Early detection of enamel demineralization by optical coherence tomography. *Sci. Rep.* **2019**, *9*, 17154. [CrossRef]
82. Wijesinghe, R.E.; Cho, N.H.; Park, K.; Jeon, M.; Kim, J. Bio-Photonic Detection and Quantitative Evaluation Method for the Progression of Dental Caries Using Optical Frequency-Domain Imaging Method. *Sensors* **2016**, *16*, 2076. [CrossRef] [PubMed]
83. Natsume, Y.; Nakashima, S.; Sadr, A. Estimation of lesion progress in artificial root caries by swept source optical coherence tomography in comparison to transverse microradiography. *J. Biomed. Opt.* **2011**, *16*, 071408. [CrossRef] [PubMed]
84. Shimamura, Y.; Murayama, R.; Kurokawa, H.; Miyazaki, M.; Mihata, Y.; Kmaguchi, S. Influence of tooth-surface hydration conditions on optical coherence-tomography imaging. *J. Dent.* **2011**, *39*, 572–577. [CrossRef] [PubMed]
85. Nazari, A.; Sadr, A.; Funollet, M.C.; Nakashima, S.; Shimada, Y.; Tagami, J.; Sumi, Y. Effect of hydration on assessment of early enamel lesion using swept-source optical coherence tomography. *J. Biophotonics* **2012**, *6*, 171–177. [CrossRef] [PubMed]
86. Won, J.; Huang, P.C.; Spillman, D.R.; Chaney, E.J.; Adam, R.; Klukowska, M.; Barkalifa, R.; Boppart, S.A. Handheld optical coherence tomography for clinical assessment of dental plaque and gingiva. *J. Biomed. Opt.* **2020**, *25*, 116011. [CrossRef] [PubMed]

MDPI
St. Alban-Anlage 66
4052 Basel
Switzerland
www.mdpi.com

Applied Sciences Editorial Office
E-mail: applsci@mdpi.com
www.mdpi.com/journal/applsci